PBSAA
Practice Papers

UniAdmissions

ISBN 978-1-912557-20-2

Published by *RAR Medical Services Limited*
www.uniadmissions.co.uk
info@uniadmissions.co.uk
Tel: 0208 068 0438

PBSAA
Practice Papers

Dr Rohan Agarwal

UniAdmissions

About the Author

Rohan is the **Director of Operations** at *UniAdmissions* and is responsible for its technical and commercial arms. He graduated from Gonville and Caius College, Cambridge and is a fully qualified doctor. Over the last five years, he has tutored hundreds of successful Oxbridge and Medical applicants. He has also authored ten books on admissions tests and interviews.

Rohan has taught physiology to undergraduates and interviewed medical school applicants for Cambridge. He has published research on bone physiology and writes education articles for the Independent and Huffington Post. In his spare time, Rohan enjoys playing the piano and table tennis.

Introduction

The Basics

The Psycological and Behavioural Sciences Admissions Assessment is an aptitude test taken by students who are applying to study Psychological and Behavioural Sciences at Cambridge. The PBSAA consists of two sections.

ONE	Problem-solving skills, including numerical and spatial reasoning. Critical thinking skills, including understanding argument and reasoning using everyday language.	ver 2 of 3 section (Thinking Skill:)ulsory. . either Par hematics and Bi	80 minutes
TWO	Ability to organise ideas in a clear and concise manner, and communicate them effectively in writing. Questions are usually)ut not necessarily medical.	essay from a chc	40 minutes

Preparing for the PBSAA

Before going any further, it's important that you understand the optimal way to prepare for the PBSAA. Rather than jumping straight into doing mock papers, it's essential that you start by understanding the components and the theory behind the PBSAA by using a PBSAA textbook. Once you've finished the non-timed practice questions, you can progress to past PBSAA papers. These are freely available online at www.uniadmissions.co.uk/PBSAA-past-papers and serve as excellent practice. Finally, once you've exhausted past papers, move onto the mock papers in this book.

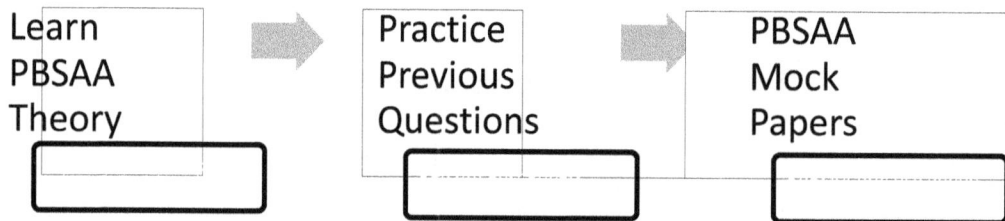

Learn PBSAA Theory   →   Practice Previous Questions   →   PBSAA Mock Papers

Already seen them all?

So, you've run out of past papers? Well hopefully that is where this book comes in. It contains two unique mock papers written by our expert oxbridge tutors.

Having successfully gained a place on their course of choice, our tutors are intimately familiar with the PBSAA and its associated admission procedures. So, the novel questions presented to you here are of the correct style and difficulty to continue your revision and stretch you to meet the demands of the PBSAA.

General Advice

Start Early

It is much easier to prepare if you practice little and often. Start your preparation well in advance; ideally 10 weeks but at the latest within a month. This way you will have plenty of time to complete as many papers as you wish to feel comfortable and won't have to panic and cram just before the test, which is a much less effective and more stressful way to learn. In general, an early start will give you the opportunity to identify the complex issues and work at your own pace.

Prioritise

The MCQ section can be very time-pressured, and if you fail to answer the questions within the time limit you will be doing yourself a major disservice as every mark counts for this section. You need to be aware of how much time you're spending on each passage and allocate your time wisely. For example, since there are 22 questions in Section 1, and you are given 40 minutes in total, you will ideally take about 100 seconds per question (including reading time) so that you will not run out of time and panic towards the end.

Positive Marking

There are no penalties for incorrect answers; you will gain one for each right answer and will not get one for each wrong or unanswered one. This provides you with the luxury that you can always guess should you absolutely be not able to figure out the right answer for a question or run behind time. Since each question in Section 1 provides you with 5 possible answers, you have a 20% chance of guessing correctly. Therefore, if you aren't sure (and are running short of time), then make an educated guess and move on. Before 'guessing' you should try to eliminate a couple of answers to increase your chances of getting the question correct. For example, if a question has 5 options and you manage to eliminate 2 options- your chances of getting the question increase from 20% to 33%!

Avoid losing easy marks on other questions because of poor exam technique. Similarly, if you have failed to finish the exam, take the last ten seconds to guess the remaining questions to at least give yourself a chance of getting them right.

Practice

This is the best way of familiarising yourself with the style of questions and the timing for this section. Although the test does not demand any technical knowledge, you are unlikely to be familiar with the style of questions in all sections when you first encounter them. Therefore, you want to be comfortable at using this before you sit the test.

Practising questions will put you at ease and make you more comfortable with the exam. The more comfortable you are, the less you will panic on the test day and the more likely you are to score highly. Initially, work through the questions at your own pace, and spend time carefully reading the questions and looking at any additional data. When it becomes closer to the test, **make sure you practice the questions under exam conditions**.

Past Papers

Official past papers and answers are freely available online at **www.uniadmissions.co.uk/PBSAA-past-papers**. Practice makes perfect, and the more you practice the questions, especially for Section 1, the better you will get. Do not worry if you make plenty of mistakes at the start, the best way to learn is to understand why you have made certain mistakes and to not commit them again in the future!

Repeat Questions

When checking through answers, pay particular attention to questions you have got wrong. If there is a worked answer, look through that carefully until you feel confident that you understand the reasoning, and then repeat the question without help to check that you can do it. If only the answer is given, have another look at the question and try to work out why that answer is correct. This is the best way to learn from your mistakes, and means you are less likely to make similar mistakes when it comes to the test. The same applies for questions which you were unsure of and made an educated guess which was correct, even if you got it right. When working through this book, **make sure you highlight any questions you are unsure of**, this means you know to spend more time looking over them once marked.

No Calculators and Dictionaries

The PBSAA requires a strong command of the English language, especially for Section A + C. You are not allowed to use spell check or a dictionary, hence you should ensure that you written English is up to standard and you should ideally make close to no grammatical or spelling errors for your essay.

Section A contains several numerical reasoning questions, and you are not allowed to use a calculator, so make sure you are careful with your calculations.

Keywords

If you're stuck on a question, sometimes you can simply quickly scan the passage for any keywords that match the questions.

A word on Timing...

"If you had all day to do your exam, you would get 100%. But you don't."

Whilst this isn't completely true, it illustrates a very important point. Once you've practiced and know how to answer the questions, the clock is your biggest enemy. This seemingly obvious statement has one very important consequence. **The way to improve your score is to improve your speed.** There is no magic bullet. But there are a great number of techniques that, with practice, will give you significant time gains, allowing you to answer more questions and score more marks.

Timing is tight throughout – **mastering timing is the first key to success**. Some candidates choose to work as quickly as possible to save up time at the end to check back, but this is generally not the best way to do it. Often questions can have a lot of information in them – each time you start answering a question it takes time to get familiar with the instructions and information. By splitting the question into two sessions (the first run-through and the return-to-check) you double the amount of time you spend on familiarising yourself with the data, as you have to do it twice instead of only once. This costs valuable time. In addition, candidates who do check back may spend 2–3 minutes doing so and yet not make any actual changes. Whilst this can be reassuring, it is a false reassurance as it is unlikely to have a significant effect on your actual score. Therefore, it is usually best to pace yourself very steadily, aiming to spend the same amount of time on each question and finish the final question in a section just as time runs out. This reduces the time spent on re-familiarising with questions and maximises the time spent on the first attempt, gaining more marks.

It is essential that you don't get stuck with the hardest questions – no doubt there will be some. In the time spent answering only one of these you may miss out on answering three easier questions. If a question is taking too long, choose a sensible answer and move on. Never see this as giving up or in any way failing, rather it is the smart way to approach a test with a tight time limit. With practice and discipline, you can get very good at this and learn to maximise your efficiency. It is not about being a hero and aiming for full marks – this is almost impossible and very much unnecessary. It is about maximising your efficiency and gaining the maximum possible number of marks within the time you have.

Use the Options:

Some passages may try to trick you by providing a lot of unnecessary information. When presented with long passages that are seemingly hard to understand, it's essential you look at the answer options so you can focus your mind. This can allow you to reach the correct answer a lot more quickly. Consider the example below:

'Mountain climbing is viewed by some as an extreme sport, while for others it is simply an exhilarating pastime that offers the ultimate challenge of strength, endurance, and sacrifice. It can be highly dangerous, even fatal, especially when the climber is out of his or her depth, or simply gets overwhelmed by weather, terrain, ice, or other dangers of the mountain. Inexperience, poor planning, and inadequate equipment can all contribute to injury or death, so knowing what to do right matters.

Despite all the negatives, when done right, mountain climbing is an exciting, exhilarating, and rewarding experience. This article is an overview beginner's guide and outlines the initial basics to learn. Each step is deserving of an article in its own right, and entire tomes have been written on climbing mountains, so you're advised to spend a good deal of your beginner's learning immersed in reading widely. This basic overview will give you an idea of what is involved in a climb.'

Looking at the options first makes it obvious that certain information is redundant and allows you to quickly zoom in on certain keywords you should pick up on in order to answer the questions.

In other cases, **you may actually be able to solve the question without having to read the passage over and over again**. For example:

Which statement best summarises this paragraph?
A. Mountain climbing is an extreme sport fraught with dangers.
B. Without extensive experience embarking on a mountain climb is fatal.
C. A comprehensive literature search is the key to enjoying mountain climbing.
D. Mountain climbing is difficult and is a skill that matures with age if pursued.
E. The terrain is the biggest unknown when climbing a mountain and therefore presents the biggest danger.

If you read the passage first before looking at the question, you might have forgotten what the passage mentioned, and you will have to spend extra time going back to the passage to re-read it again.

You can **save a lot of time by looking at the questions first before reading the passage**. After looking at the question, you will know at the back of your head to look out for and this will save a considerable amount of time.

Manage your Time:

It is highly likely that you will be juggling your revision alongside your normal school studies. Whilst it is tempting to put your A-levels on the back burner falling behind in your school subjects is not a good idea, don't forget that to meet the conditions of your offer should you get one you will need at least one A*. So, time management is key!

Make sure you set aside a dedicated 90 minutes (and much more closer to the exam) to commit to your revision each day. The key here is not to sacrifice too many of your extracurricular activities, everybody needs some down time, but instead to be efficient. Take a look at our list of top tips for increasing revision efficiency below:

1. Create a comfortable work station
2. Declutter and stay tidy
3. Treat yourself to some nice stationery
4. See if music works for you → if not, find somewhere peaceful and quiet to work
5. Turn off your mobile or at least put it into silent mode
6. Silence social media alerts
7. Keep the TV off and out of sight
8. Stay organised with to do lists and revision timetables – more importantly, stick to them!
9. Keep to your set study times and don't bite off more than you can chew
10. Study while you're commuting
11. Adopt a positive mental attitude
12. Get into a routine
13. Consider forming a study group to focus on the harder exam concepts
14. Plan rest and reward days into your timetable – these are excellent incentive for you to stay on track with your study plans!

Keep Fit & Eat Well:

'A car won't work if you fill it with the wrong fuel' - your body is exactly the same. You cannot hope to perform unless you remain fit and well. The best way to do this is not underestimate the importance of healthy eating. Beige, starchy foods will make you sluggish; instead start the day with a hearty breakfast like porridge. Aim for the recommended 'five a day' intake of fruit/veg and stock up on the oily fish or blueberries – the so called "super foods".

When hitting the books, it's essential to keep your brain hydrated. If you get dehydrated you'll find yourself lethargic and possibly developing a headache, neither of which will do any favours for your revision. Invest in a good water bottle that you know the total volume of and keep sipping throughout the day. Don't forget that the amount of water you should be aiming to drink varies depending on your mass, so calculate your own personal recommended intake as follows: 30 ml per kg per day.

It is well known that exercise boosts your wellbeing and instils a sense of discipline. All of which will reflect well in your revision. It's well worth devoting half an hour a day to some exercise, get your heart rate up, break a sweat, and get those endorphins flowing.

Sleep

It's no secret that when revising you need to keep well rested. Don't be tempted to stay up late revising as sleep actually plays an important part in consolidating long term memory. Instead aim for a minimum of 7 hours good sleep each night, in a dark room without any glow from electronic appliances. Install flux (https://justgetflux.com) on your laptop to prevent your computer from disrupting your circadian rhythm. Aim to go to bed the same time each night and no hitting snooze on the alarm clock in the morning!

Revision Timetable

Still struggling to get organised? Then try filling in the example revision timetable below, remember to factor in enough time for short breaks, and stick to it! Remember to schedule in several breaks throughout the day and actually use them to do something you enjoy e.g. TV, reading, YouTube etc.

MOND	
TUESD	
WEDNES	
THURSI	
FRIDAY	
SATURI	
SUNDAY	
EXAMI DAY	School 1a 1b 2

Top tip! Ensure that you take a watch that can show you the time in seconds into the exam. This wi allow you have a much more accurate idea of the time you're spending on a question. In general, if you'v spent >150 seconds on a section 1 question – move on regardless of how close you think you are to solvir

Getting the most out of Mock Papers

Mock exams can prove invaluable if tackled correctly. Not only do they encourage you to start revision earlier, they also allow you to **practice and perfect your revision technique**. They are often the best way of improving your knowledge base or reinforcing what you have learnt. Probably the best reason for attempting mock papers is to familiarise yourself with the exam conditions of the PBSAA as they are particularly tough.

Start Revision Earlier

Thirty five percent of students agree that they procrastinate to a degree that is detrimental to their exam performance. This is partly explained by the fact that they often seem a long way in the future. In the scientific literature this is well recognised, Dr. Piers Steel, an expert on the field of motivation states that *'the further away an event is, the less impact it has on your decisions'*.

Mock exams are therefore a way of giving you a target to work towards and motivate you in the run up to the real thing – every time you do one treat it as the real deal! If you do well then it's a reassuring sign; if you do poorly then it will motivate you to work harder (and earlier!).

Practice and perfect revision techniques

In case you haven't realised already, revision is a skill all to itself, and can take some time to learn. For example, the most common revision techniques including **highlighting and/or re-reading are quite ineffective** ways of committing things to memory. Unless you are thinking critically about something you are much less likely to remember it or indeed understand it.

Mock exams, therefore allow you to test your revision strategies as you go along. Try spacing out your revision sessions so you have time to forget what you have learnt inbetween. This may sound counterintuitive but the second time you remember it for longer. Try teaching another student what you have learnt, this forces you to structure the information in a logical way that may aid memory. Always try to question what you have learnt and appraise its validity. Not only does this aid memory but it is also a useful skill for the PBSAA, Cambridge interviews, and beyond.

Improve your knowledge

The act of applying what you have learnt reinforces that piece of knowledge. An essay question in Section 2 may ask you about a fairly simple topic, but if you have a deep understanding of it you are able to write a critical essay that stands out from the crowd. Essay questions in particular provide a lot of room for students who have done their research to stand out, hence you should always aim to improve your knowledge and apply it from time to time. As you go through the mocks or past papers take note of your performance and see if you consistently under-perform in specific areas, thus highlighting areas for future study.

Get familiar with exam conditions

Pressure can cause all sorts of trouble for even the most brilliant students. The PBSAA is a particularly time pressured exam with high stakes – your future (without exaggerating) does depend on your result to a great extent. The real key to the PBSAA is overcoming this pressure and remaining calm to allow you to think efficiently.

Mock exams are therefore an excellent opportunity to devise and perfect your own exam techniques to beat the pressure and meet the demands of the exam. **Don't treat mock exams like practice questions – it's imperative you do them under time conditions.**

Before using this Book

Do the ground work

➢ Understand the format of the PBSAA – have a look at the PBSAA website and familiarise yourself with it: https://www.undergraduate.study.cam.ac.uk/courses/psychological-and-behavioural-sciences

➢ Read widely in order to prepare yourself for Section 2.

➢ Improve your written English if you are not confident in this aspect by practicing writing and reading frequently.

➢ Try to broaden your reading by learning about different topics that you are unfamiliar with as the essay topics can vary greatly.

➢ Learn how to understand a writer's viewpoint by reading news articles and having a go at summarising what the writer is arguing about.

➢ Be consistent – slot in regular PBSAA practice sessions when you have pockets of free time.

➢ Engage in discussion sessions with your friends and teachers – this might give you more ideas about certain essay topics.

Ease in gently

With the ground work laid, there's still no point in adopting exam conditions straight away. Instead invest in a beginner's guide to the PBSAA, which will not only describe in detail the background and theory of the exam, but take you through section by section what is expected. *The Ultimate PBSAA Guide* is the most popular PBSAA textbook – you can get a free copy by flicking to the back of this book.

When you are ready to move on to past papers, take your time and puzzle your way through all the questions. Really try to understand solutions. A past paper question won't be repeated in your real exam, so don't rote learn methods or facts. Instead, focus on applying prior knowledge to formulate your own approach.

If you're really struggling and have to take a sneak peek at the answers, then practice thinking of alternative solutions, or arguments for essays. It is unlikely that your answer will be more elegant or succinct than the model answer, but it is still a good task for encouraging creativity with your thinking. Get used to thinking outside the box!

Accelerate and Intensify

Start adopting exam conditions after you've done two past papers. Don't forget that **it's the time pressure that makes the PBSAA hard** – if you had as long as you wanted to sit the exam you would probably get 100%.

Doing all the past papers is a good target for your revision. Choose a paper and proceed with strict exam conditions. Take a short break and then mark your answers before reviewing your progress. For revision purposes, as you go along, keep track of those questions that you guess – these are equally as important to review as those you get wrong.

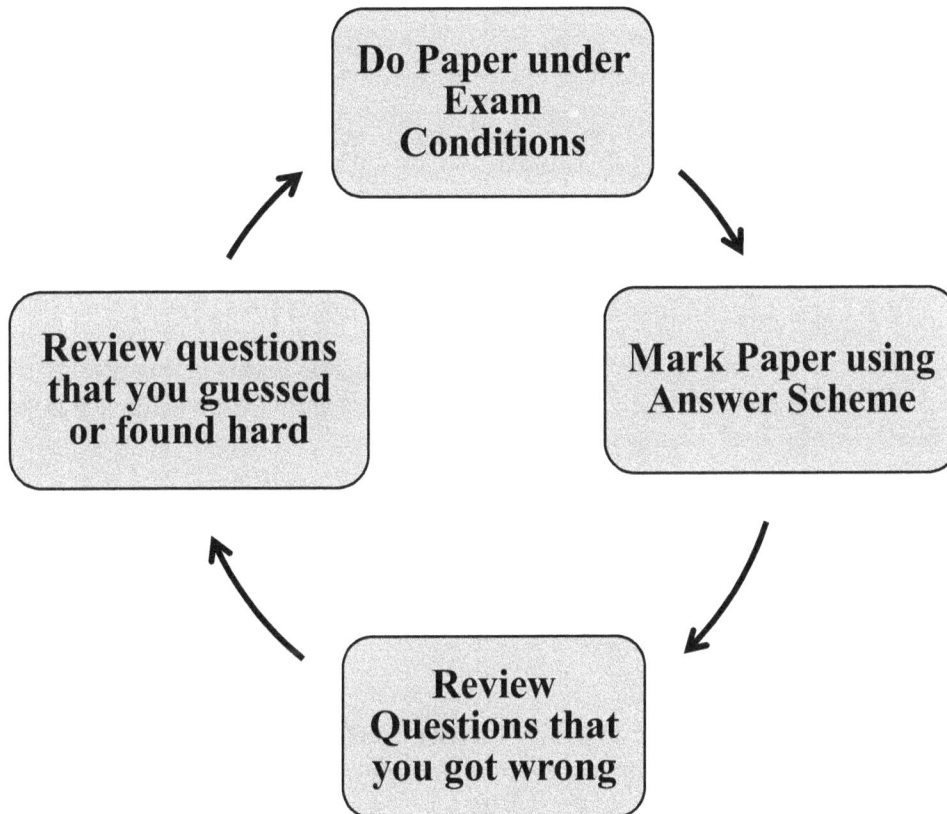

```
        ┌──────────────────┐
        │ Do Paper under   │
        │     Exam         │
        │   Conditions     │
        └──────────────────┘
   ┌──────────────────┐   ┌──────────────────┐
   │ Review questions │   │ Mark Paper using │
   │ that you guessed │   │  Answer Scheme   │
   │  or found hard   │   └──────────────────┘
   └──────────────────┘
            ┌──────────────────┐
            │     Review       │
            │ Questions that   │
            │  you got wrong   │
            └──────────────────┘
```

Once you've exhausted all the past papers, move on to tackling the unique mock papers in this book. In general, you should aim to complete one to two mock papers every night in the ten days preceding your exam.

Section 1: An Overview

What will you be tested on?	Questions	Duration
Problem-solving skills, numerical and spatial reasoning, critical thinking skills, understanding arguments and reasoning	Answer 2 of 3 sections; the first (Part A Thinking Skills) is compulsory. Then answer either Part B Mathematics and Biology or Part C Reading Comprehension.	80 Minutes

This is the first section of the PBSAA. You have 80 minutes in total to complete the MCQ questions, including reading time. Part A (Thinking Skillss) is compulsory consisting of 22 MCQ questions, 11 will be critical thinking questions, 11 will be problem solving questions. Part B (Mathematics and Biology) consists of 30 MCQ questions, 15 for Mathematics and 15 for Biology. These require the candidate to use and apply mathematical and biological knowledge. Part C (Reading Comprehension) consists of 24 MCQ questions and comprises three tasks.

Not all the questions are of equal difficulty and so as you work through the past material it is certainly worth learning to recognise quickly which questions you should spend less time on in order to give yourself more time for the trickier questions.

Deducing arguments

Several MCQ questions will be aimed at testing your understand of the writer's argument. It is common to see questions asking you 'what is the writer's view?' or 'what is the writer trying to argue?'. This is arguably an important skill you will have to develop, and the PBSAA is designed to test this ability. You have limited time to read the passage and understand the writer's argument, and the only way to improve your reading comprehension skill is to read several well-written news articles on a daily basis and think about them in a critical manner.

Assumptions

It is important to be able to identify the assumptions that a writer makes in the passage, as several questions might question your understand of what is assumed in the passage. For example, if a writer mentions that 'if all else remains the same, we can expect our economic growth to improve next year', you can identify an assumption being made here – the writer is clearly assuming that all external factors remain the same.

Fact vs Opinion

It is important to **be able to decipher whether the writer is stating a fact or an opinion** – the distinction is usually rather subtle, and you will have to decide whether the writer is giving his or her own personal opinion, or presenting something as a fact. Section 1 may contain questions that will test your ability to identify what is presented as a fact and what is presented as an opinion.

Fact	Opinion
'There are 7 billion people in this world…'	'I believe there are more than 7 billion people in this world…'
'She is an Australian…'	'She sounded like an Australian…'
'Trump is the current President…'	'Trump is a horrible President…'
'Vegetables contain a lot of fibre…'	'Vegetables are good for you…'

Numerical and spatial reasoning

There are several questions that will test how well you can cope with numbers, and you should ideally be comfortable with simple mental calculations and being able to think logically.

Section 2: An Overview

What will you be tested on?	Questions	Duration
Your ability to discuss a quote under timed conditions, your writing technique and your argumentative abilities	One out of a choice of four	40 Minutes

Section 2 is usually what students are more comfortable with – after all, many GCSE and A Level subjects require you to write essays within timed conditions. This section does not require you to have any particular specialist knowledge – the questions can be very broad and cover a wide range of topics. The question will be focused around a quotation, which is related to various different topics.

Here are some of the topics that might be linked to quotations in Section 2:

➢ Science	➢ Religion	➢ Ethics	➢ Philosophy	➢ History
➢ Politics	➢ Technology	➢ Morality	➢ Education	➢ Geopolitics

As you can see, this list is very broad and definitely non-exhaustive, and you do not get many choices to choose from (you have to write one essay out of four choices). Many students make the mistake of focusing too narrowly on one or two topics that they are comfortable with – this is a dangerous gamble and if you end up with four questions you are unfamiliar with, this is likely to negatively impact your score. **You should ideally focus on three topics to prepare from the list above**, and you can pick which topics from the list above are the ones you would be more interested in. Here are some suggestions:

Science

An essay that is related to science might related to recent technological advancements and their implications, such as the rise of Bitcoin and the use of blockchain technology and artificial intelligence. This is interrelated to ethical and moral issues; hence you cannot merely just regurgitate what you know about artificial intelligence or blockchain technology. **The examiners do not expect you to be an expert in an area of science** – what they want to see is how you identify certain moral or ethical issues that might arise due to scientific advancements, and how do we resolve such conundrums as human beings.

Politics

Politics is undeniably always a hot topic and consequently a popular choice amongst students. The danger with writing a politics question is that some **students get carried away and make their essay too one-sided or emotive** – for example a student may chance upon an essay question related to Brexit and go on a long rant about why the referendum was a bad idea. You should always remember to answer the question and make sure your essay addresses the exact question asked – do not get carried away and end up writing something irrelevant just because you have strong feelings about a certain topic.

Religion

Religion is always a controversial issue and essays on religion provide good students with an excellent opportunity to stand out and display their maturity in thought. Questions can range from asking about your opinion with regards to banning the wearing of a headdress to whether children should be exposed to religious practices at a young age. Questions related to religion will **require you to be sensitive and measured** in your answers - it is easy to trip up on such questions if you're not careful.

Education

Education is perhaps always a relatable topic to students, and students can draw from their own experience with the education system in order to form their opinion and write good essays on such topics. Questions can range from whether university places should be reduced, to whether we should be focusing on learning the sciences as opposed to the arts.

Section 2: Revision Guide

Science

	Resource	What to read/do
1.	Newspaper Articles	• The Guardian, The Times, The Economist, The Financial Times, The Telegraph, The New York Times, The Independent
2.	A Levels/IB	• Look at the content of your science A Levels/IB if you are doing science subjects and critically analyse what are the potential moral/ethical implications • Use your A Levels/IB resources in order to seek out further readings – e.g. links to a scientific journal or blog commentary
3.	Online videos	• There are plenty of free resources online that provide interesting commentary on science and the moral and ethical conundrums that scientists face on a daily basis • E.g. Documentaries and specialist science channels on YouTube • National Geographic, Animal Planet etc. might also be good if you have access to them
4.	Debates	• Having a discussion with your friends about topics related to science might also help you formulate some ideas • Attending debate sessions where the topic is related to science might also provide you with excellent arguments and counter-arguments • Some universities might also host information sessions for sixth form students – some might be relevant to ethical and moral issues in science
5.	Museums	• Certain museums such as the Natural Science Museum might provide some interesting information that you might not have known about
6.	Non-fiction books	• There are plenty of non-fiction books (non-technical ones) that might discuss moral and ethical issues about science in an easily digestible way

Politics

	Resource	What to read/do
1.	Newspaper Articles	• The Guardian, The Times, The Economist, The Financial Times, The Telegraph, The New York Times, The Independent
2.	Television	• Parliamentary sessions • Prime Minister Questions • Political news
3.	Online videos	• Documentaries • YouTube Channels
4.	Lectures	• University introductory lectures • Sixth form information sessions
5.	Debates	• Debates held in school • Joining a politics club
6.	Podcasts	• Political podcasts • Listen to both sides to get a more rounded view (e.g. listening to both left and right-wing podcasts)

Religion

Syllabus I	What to read/do
2. Non-fi books	• Read up about books that explain the origins and beliefs of different typ religion • E.g. Books that talk about the origins of Christianity, Islam or Buddhism, the books etc.
3. Talkin religio leader:	religions more and being able to write an essay on religion with more maturity nuance • Talking to people from different religious backgrounds may also be a good w forming a more well-rounded opinion
5. Lectur	• Information sessions • Relevant introductory lectures

Education

	What to read/do
1. Newspaper Articles	• The Guardian, The Times, The Economist, The Financial Times, T Telegraph, The New York Times, The Independent
2. A Levels/IB	you feel like what you are studying is useful and relevant? E.g. Studying a versus science • Compare the education you are receiving with your friends in differ schools or different subjects
4. University applications	• Have a read of how different universities promote themselves – do th claim to provide students with academic enlightenment, or better prospects, or a good social life? • Why do different universities focus on different things?
6. Talk to your teachers	• Your teachers have been in the education industry for years and may decades – talk to them and ask them for their opinion • Talk to different teachers and compare their opinions regarding how

Top Tip! Although you aren't required to have extra knowledge for the PBSAA essay, doing so will allow you to make your essay stand out from the crowd. However, you should first prioritise perfecting your writing style rather than doing extra reading as the former will have a greater impact on your mark.

How to use this Book

If you have done everything this book has described so far then you should be well equipped to meet the demands of the PBSAA, and therefore **the mock papers in the rest of this book should ONLY be completed under exam conditions**.

This means:
➢ Absolute silence – no TV or music
➢ Absolute focus – no distractions such as eating your dinner
➢ Strict time constraints – no pausing half way through
➢ No checking the answers as you go
➢ Give yourself a maximum of three minutes between sections – keep the pressure up
➢ Complete the entire paper before marking
➢ Mark harshly

In practice this means setting aside 80 minutes for Section 1 and 40 minutes for Section 2 in an evening to find a quiet spot without interruptions and tackle the paper. Completing one mock paper every evening in the week running up to the exam would be an ideal target.

➢ Tackle the paper as you would in the exam.
➢ Return to mark your answers, but mark harshly if there's any ambiguity.
➢ Highlight any areas of concern.
➢ If warranted read up on the areas you felt you underperformed to reinforce your knowledge.
➢ If you inadvertently learnt anything new by muddling through a question, go and tell somebody about it to reinforce what you've discovered.

Finally relax… the PBSAA is an exhausting exam, concentrating so hard continually for 1.5 hours will take its toll. So, being able to relax and switch off is essential to keep yourself sharp for exam day! Make sure you reward yourself after you finish marking your exam.

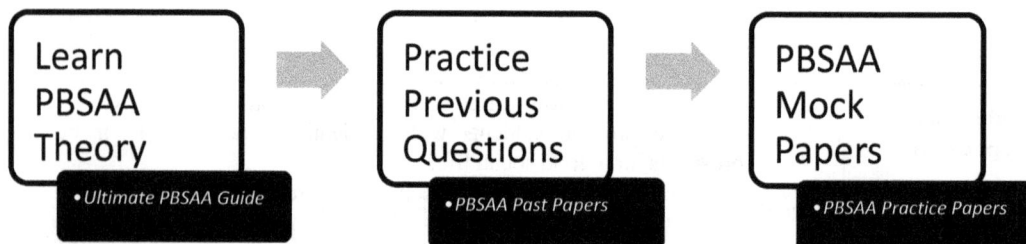

Learn PBSAA Theory	→	Practice Previous Questions	→	PBSAA Mock Papers
•*Ultimate PBSAA Guide*		•*PBSAA Past Papers*		•*PBSAA Practice Papers*

Scoring Tables

Use these to keep a record of your scores from past papers – you can then easily see which paper you should attempt next (always the one with the lowest score).

	3rd Attempt
Specimen	
Mock Paper	

	3rd Attempt
Specimen	
Mock Paper	

	3rd Attempt
Specimen	
2016	
Mock Paper	

	3rd Attempt
2016	
Mock Paper	

You will not be able to give yourself a score for Section 2– the best way to gauge your performance for Section 2 will be to compare your arguments and counter-arguments with the model answer, or let your friends or teachers read it and gather some feedback from them. Fortunately for the mock papers in this book, there are model answers for you to compare your essays against!

Mock Paper A

Section 1A

Question 1

Competitors need to be able to run 200 metres in under 25 seconds to qualify for a tournament. James, Steven and Joe are attempting to qualify. Steven and Joe run faster than James. James' best time over 200 metres is 26.2 seconds. Which response is definitely true?

A. Only Joe qualifies

B. James does not qualify.

C. Joe and Steven both qualify

D. Joe qualifies

E. No one qualifies

Question 2

You spend £5.60 in total on a sandwich, a packet of crisps and a watermelon. The watermelon cost twice as much as the sandwich, and the sandwich cost twice the price of the crisps. How much did the watermelon cost?

A. £1.20

B. £2.60

C. £2.80

D. £3.20

E. £3.60

Question 3

Jane, Chloe and Sam are all going by train to a football match. Chloe gets the 2:15pm train. Sam's journey takes twice as long Jane's. Sam catches the 3:00pm train. Jane leaves 20 minutes after Chloe and arrives at 3:25pm. When will Sam arrive?

A. 3:50pm

B. 4:10pm

C. 4:15pm

D. 4:30pm

E. 4:40pm

Question 4

Michael has eleven sweets. He gives three sweets to Hannah. Hannah now has twice the number of sweets Michael has remaining. How many sweets did Hannah have before the transaction?

A. 11

B. 12

C. 13

D. 14

E. 15

Question 5

Alex gets a pay rise of 5% plus an extra £6 per week. The flat rate of income tax is decreased from 14% to 12% at the same time. Alex's current weekly take-home pay is £250 per week.
What will his new weekly take-home pay be, to the nearest whole pound?

A. £260

B. £267

C. £273

D. £279

E. £285

Question 6

You have four boxes, each containing two coloured cubes. Box A contains two white cubes, Box B contains two black cubes, and Boxes C and D both contain one white cube and one black cube. You pick a box at random and take out one cube. It is a white cube. You then draw another cube from the same box.
What is the probability that this cube is not white?

A. ½

B. ⅓

C. ⅔

D. ¼

E. ¾

Question 7

Anderson & Co. hire out heavy plant machinery at a cost of £500 per day. There is a surcharge for heavy usage, at a rate of £10 per minute of usage over 80 minutes. Concordia & Co. charge £600 per day for similar machinery, plus £5 for every minute of usage.

For what duration of usage are the costs the same for both companies?

A. 100 minutes   B. 130 minutes   C. 140 minutes   D. 170 minutes   E. 180 minutes

Question 8

Simon is discussing with Seth whether or not a candidate is suitable for a job. When pressed for a weakness at interview, the candidate told Simon that he is a slow eater. Simon argues that this will reduce the candidate's productivity, since he will be inclined to take longer lunch breaks.

Which statement **best** substantiates Simon's argument?

A. Slow eaters will take longer to eat lunch
B. Longer lunch breaks are a distraction
C. Eating more slowly will reduce the time available to work
D. Eating slowly is a weakness
E. People who like food are more likely to eat slowly

Question 9

Three pieces of music are on repeat in different rooms of a house. One piece of music is three minutes long, one is four minutes long and the final one is 100 seconds long. All pieces of music start playing at exactly the same time. How long is it until they are next all starting together?

A. 12 minutes   B. 15 minutes   C. 20 minutes   D. 60 minutes   E. 300 minutes

Question 10

A car leaves Salisbury at 8:22am and travels 180 miles to Lincoln, arriving at 12:07pm. Near Warwick, the driver stopped for a 14-minute break. What was its average speed, whilst travelling, in kilometres per hour? It should be assumed that the conversion from miles to kilometres is 1:1.6.

A. 51kph   B. 67kph   C. 77kph   D. 82kph   E. 386kph

Questions **11** and **12** refer to the following data:

Five respondents were asked to estimate the value of three bottles of wine, in pounds sterling.

Respondent	Wine 1	Wine 2	Wine 3
1	13	16	25
2	17	16	23
3	11	17	21
4	13	15	14
5	15	19	29
Actual retail value	8	25	23

Question 11

What is the mean error margin in the guessing of the value of wine 1?

A. £4.80　　　　B. £5.60　　　　C. £5.80　　　　D. £6.20　　　　E. £6.40

Question 12

Which respondent guessed most accurately on average?

A. Respondent 1　　B. Respondent 2　　C. Respondent 3　　D. Respondent 4　　E. Respondent 5

Questions **13** and **14** refer to the following data:

The population of Country A is 40% greater than the population of Country B.

The population of Country C is 30% less than the population of Country D (which is has a population 20% greater than Country B).

Question 13

Given that the population of Country A is 45 million, what is the population of country D?

A. 32.1 million　　B. 35.8 million　　C. 36.6 million　　D. 38.6 million　　E. 39.0 million

Question 14

The population of Country A is still 45 million. If Country B introduced a new health initiative costing $45 per capita, what would be the total cost?

A. $1.35 bn　　B. $1.45 bn　　C. $1.50 bn　　D. $1.55 bn　　E. $1.65 bn

Question 15

A car averages a speed of 30mph over a certain distance and then returns over the same distance at an average speed of 20mph. What is the average speed for the journey as a whole?

A. 22.5 mph　　B. 24 mph　　C. 25 mph　　D. 26 mph

E. The distance travelled is required to calculate average speed

Question 16

All sheep are ruminants and all marsupials are mammals. No sheep are marsupials. Which of the following must be true?

A. Some ruminants are marsupials.　　B. All mammals are marsupials　　C. All sheep are mammals

D. Some sheep are marsupials.　　E. None of the above

Question 17

The price of toothpaste rises by 80%. This is later reduced by 50% due to competition. Zoe buys two tubes of toothpaste and gets the third free because of a loyalty card. How much did she have to pay per tube of toothpaste? Express your answer as a percentage of the original price.

A. 16.67%　　B. 33%　　C. 60%　　D. 66.7%　　E. 100%

Question 18

Reports of cybercrime are increasing year on year. Last year, police dealt with 250% more cybercrime then the year before. Common complaints relate to inappropriate or defamatory use of social media. To deal with this, many police forces are creating dedicated teams to deal with online offences. A pilot study showed that a dedicated cybercrime team solved cases of cybercrime 40% faster than regular detectives. Therefore, the measure will act to suppress the rise in cybercrime.

Which statement best validates the above argument?

A. Solving crimes faster is necessary to keep pace with the increase in crime
B. Solving crimes faster leads to more convictions
C. Solving crimes faster increases police resources to tackle crime
D. Solving crimes faster saves money
E. Solving crimes faster reassures the public of action

Question 19

Recently in Kansas, a number of farm animals have been found killed in the fields. The nature of the injuries is mysterious, but consistent with tales of alien activity. Local people talk of a number of UFO sightings, and claim extra-terrestrial responsibility. Official investigations into these claims have dismissed them, offering rational explanations for the reported phenomena. However, these official investigations have failed to deal with the point that, even if the UFO sightings can be explained in rational terms, the injuries on the carcasses of the farm animals cannot be. Extra-terrestrial beings must therefore be responsible for these attacks."
Which of the following best expresses the main conclusion of this argument?

A. Sightings of UFOs cannot be explained by rational means
B. Recent attacks must have been carried out by extra-terrestrial beings
C. The injuries on the carcasses are not due to normal predators
D. UFO sightings are common in Kansas
E. Official investigations were a cover-up

Question 20

"To make a cake you must prepare the ingredients and then bake it in the oven. You purchase the required ingredients from the shop, however the oven is broken. Therefore, you cannot make a cake."

Which of the following arguments has the same structure?

A. To get a good job, you must have a strong CV then impress the recruiter at interview. Your CV was not as good as other applicants; therefore, you didn't get the job.
B. To get to Paris, you must either fly or take the Eurostar. There are flight delays due to dense fog, therefore you must take the Eurostar.
C. To borrow a library book, you must go to the library and show your library card. At the library, you realise you have forgotten your library card. Therefore, you cannot borrow a book.
D. To clean a bedroom window, you need a ladder and a hosepipe. Since you don't have the right equipment, you cannot clean the window.
E. Bears eat both fruit and fish. The river is frozen, so the bear cannot eat fish.

Question 21

Growing vegetables requires patience, skill and experience. Patience and skill without experience is common – but often such people give up prematurely as skill alone is insufficient to grow vegetables, and patience can quickly be exhausted.

Which of the following summarises the main argument?

A.   Most people lack the skill needed to grow vegetables
B.   Growing vegetables requires experience
C.   The most important thing is to get experience
D.   Most people grow vegetables for a short time but give up due to a lack of skill
E.   Successful vegetable growers need to have several positive traits

Question 22

Joseph has a bag of building blocks of various shapes and colours. Some of the cubic ones are black. Some of the black ones are pyramid shaped. All blue ones are cylindrical. There is a green one of each shape. There are some pink shapes.

Which of the following is definitely **NOT** true?

A.   Joseph has pink cylindrical blocks
B.   Joseph doesn't have pink cylindrical blocks
C.   Joseph has blue cubic blocks
D.   Joseph has a green pyramid
E.   Joseph doesn't have a black sphere

END OF SECTION

Section 1B

Question 1

Which of the following is NOT present in the Bowman's capsule?

A. Urea   B. Glucose   C. Sodium   D. Water   E. Haemoglobin

Question 2

The primary ions responsible for an action potential on a muscle cell membrane are Sodium and Potassium. Sodium concentration is higher than that potassium outside the cell. Potassium concentration is higher than sodium inside the cell. A muscle cell membrane depolarises. Which of the following must be true?

A   More potassium moves into the nerve cell than sodium.

B   More sodium moves into the nerve cell than potassium.

C   There is no net flow of sodium or potassium ions.

D   The membrane potential becomes more negative

E   None of the above

Question 3

Calculate the radius of a sphere which has a surface area three times as great as its volume.

A.   0.5   B. 1   C. 1.5   D. 2   E. 2.5

Question 4

Which of the following in NOT a polymer?

A. Polythene   B. Collagen   C. DNA   D. Glycogen   E. Starch   F   Triglycerides

Question 5

Place the following substances in order from most to least reactive:

1	Sodium	4	Zinc
2	Potassium	5	Copper
3	Aluminium	6	Magnesium

A.   1 » 2 » 6 » 3 » 4 » 5   B.   1 » 2 » 6 » 3 » 5 » 4   C.   » 1 » 6 » 3 » 4 » 5
D.   2 » 1 » 6 » 3 » 5 » 4   E.   2 » 6 » 1 » 3 » 4 » 5

Question 6

The normal cardiac cycle has two phases, systole and diastole. During diastole, which of the following is **NOT** true?

A. The aortic valve is closed

B. The ventricles are relaxing

C. The volume of blood in the ventricles is increasing

D. The pressure in the aorta increases

E. There is blood in the ventricles

Question 7

Below is a graph showing the concentration of product over time as substrate concentration is increased.   Some enzyme inhibitors are introduced.

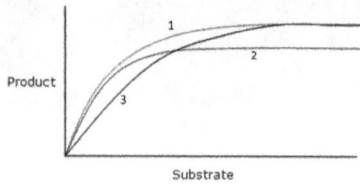

Which, if any, line represents the effect of competitive inhibition?

A.   Line 1   B.   Line 2   C.   Line 3   D.   None of these

Question 8

Which of the following is **NOT** present in the plasma membrane?
A.   Extrinsic proteins
B.   Intrinsic proteins
C.   Phospholipids
D.   Glycoproteins
E.   Nucleic Acids
F.   They are all present

Question 9

There are 1200 international airports in the world. If 4 flights take off every hour from each airport, calculate the annual number of commercial flights worldwide, to the nearest 1 million.

A.   28 million   B.   36 million   C.   42 million   D.   44 million   E.   48 million   F   56 million

Question 10

Given:
➤ $F + G + H = 1$
➤ $F + G - H = 2$
➤ $F - G - H = 3$
Calculate the value of FGH.

A.   -2   B.   -0.5   C.   0   D.   0.5   E.   2

Question 11

The concentration of chloride in the blood is 100mM. The concentration of thyroxine is 1×10^{-10}kM. Calculate the ratio of thyroxine to chloride ions in the blood.

A.   Chloride is 100,000,000 times more concentrated than thyroxine
B.   Chloride is 1,000,000 times more concentrated than thyroxine
C.   Chloride is 1000 times more concentrated than thyroxine
D.   Concentrations of chloride and thyroxine are equal
E.   Thyroxine is 1000 times more concentrated than chloride
F.   Thyroxine is 1,000,000 times more concentrated than chloride

Question 12
How many seconds are there in 66 weeks? [n! = 1 x 2 x 3 x... x n].

A.   7!   B.   8!   C.   9!   D.   10!   E.   11!   F   12!

Question 13
Which of the following is **NOT** a hormone?

A.   Insulin   B.   Glycogen   C.   Noradrenaline   D.   Cortisol   E.   Thyroxine   F   Progesterone

Question 14
Study the diagram, comprising regular pentagons. What is the product of **a** and **b**?

A.   580°   B.   1,111°   C.   3,888°
D.   7,420°   E.   9,255°

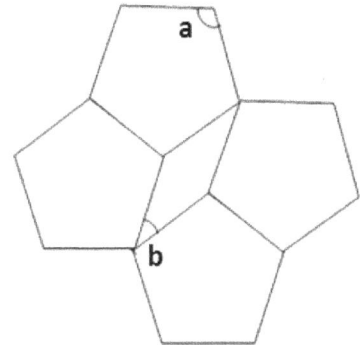

Question 15
The table below shows the results of a study investigating antibiotic resistance in staphylococcus populations.

Antibiotic	Number of Bacteria tested	Number of Resistant Bacteria
Benzyl-penicillin	10^{11}	98
Chloramphenicol	10^9	1200
Metronidazole	10^8	256
Erythromycin	10^5	2

A single staphylococcus bacterium is chosen at random from a similar population. Resistance to any one antibiotic is independent of resistance to others. Calculate the probability that the bacterium selected will be resistant to all four drugs.

A.   1 in 10^{12} B.   1 in 10^6 C.   1 in 10^{20} D.   1 in 10^{25} E.   1 in 10^{30} F 1 in 10^{35}

Question 16
Which of the following statements, regarding normal human digestion, is FALSE?

A.   Amylase is an enzyme which breaks down starch
B.   Amylase is produced by the pancreas
C.   Bile is stored in the gallbladder
D.   The small intestine is the longest part of the gut
E.   Insulin is released in response to feeding
F.   None of the above

Question 17

Jane is one mile into a marathon. Which of the following statements is **NOT** true, relative to before she started?

A. Blood flow to the skin is increased
B. Blood flow to the muscles is increased
C. Blood flow to the lungs is increased
D. Blood flow to the gut is decreased
E. Blood flow to the kidneys is decreased
F. None of the above

Question 18

A newly discovered species of beetle is found to have 29.6% Adenine (A) bases in its genome. What is the percentage of Cytosine (C) bases in the beetle's DNA?

A. 20.4%   B. 29.6%   C. 40.8%   D. 59.2%   E. 70.6%

Question 19

Study the following diagram of the human heart. What is true about structure **A**?

A. It is closed during systole
B. It prevents blood flowing into the left ventricle during systole
C. It prevents blood flowing into the right ventricle during systole
D. To prevent blood flowing into the left ventricle during diastole
E. It is open when left ventricular pressure is greater than aortic pressure
F. It is open when the right ventricle is emptying

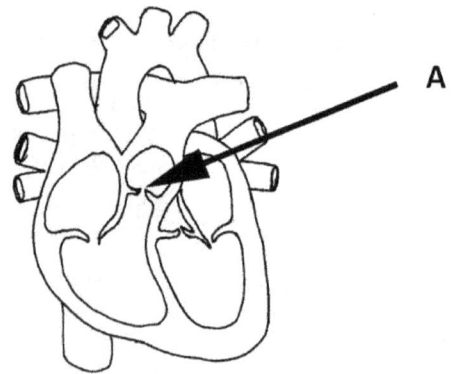

Question 20

In carbon monoxide poisoning, carbon monoxide binds irreversibly to the oxygen binding site of haemoglobin. Regarding carbon monoxide poisoning, which of the following statements is true?

A. Carbon monoxide poisoning has no serious consequences
B. Haemoglobin is heavier, as both oxygen and carbon monoxide bind to it
C. Affected individuals have a raised heart rate
D. The CO_2 carrying capacity of the blood is decreased
E. Breathing faster oxygenates the blood more at rest

Question 21

Why do cells undergo mitosis?
1. Asexual Reproduction
2. Sexual Reproduction
3. Growth of the human embryo
4. Replacement of dead cells

A.   1 only   B.   2 only   C.   3 only   D.   4 only
E.   2 and 3   F.   1,2, and 3,   G.   1,3, and 4   H.   2,3, and 4

Question 22

Calculate the perimeter of a regular polygon which has interior angles of 150° and sides of 15cm

A.   75 cm   B.   150 cm   C.   180 cm   D.   225 cm   E.   1,500 cm

Question 23

Make y the subject of the formula: $\frac{y+x}{x} = \frac{x}{a} + \frac{a}{x}$

A) $y = \frac{x^2}{a} + a$

B) $y = \frac{x^2 + a^2 - ax}{a}$

C) $y = \frac{-ax}{x^2 + a^2}$

D) $y = \frac{x^2}{ax} + a - x$

E) $y = a^2 - ax$

Question 24

The diagram below shows a series of sports fields:

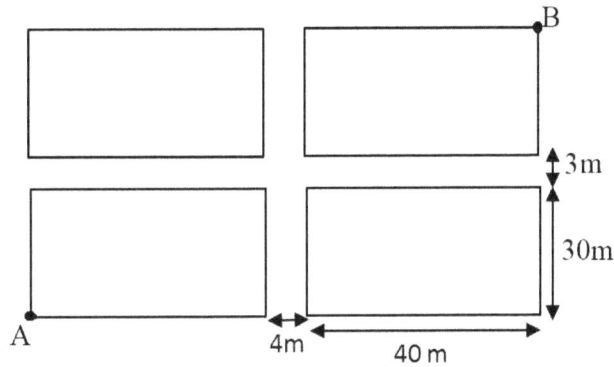

B

3m

30m

A

4m   40 m

Calculate the shortest distance between points A and B.

A.   100 m   B.   105 m   C.   146 cm   D.   148 m   E.   154 m

Question 25

Calculate $\frac{1.25 \times 10^{10} + 1.25 \times 10^9}{2.5 \times 10^8}$

A.   0   B.   1   C.   55   D.   110   E. 1.25×10^8 F. 5.5×10^7 G. 5.5×10^8

Question 26
Which row of the table is correct regarding the cell shown below?

	Most chemical reactions occur here	Involved in energy release	Cell type
A	A	B	Animal
B	A	B	Bacterial
C	A	D	Animal
D	B	D	Bacterial
E	B	B	Animal
F	B	A	Bacterial
G	D	D	Animal
H	D	B	Bacterial

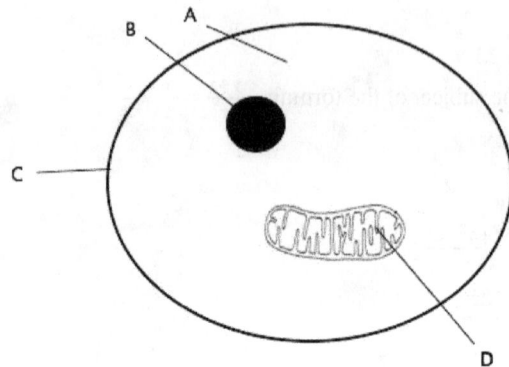

Question 27
Solve $y = 2x - 1$ and $y = x^2 - 1$ for x and y.

A. (0, -1) and (2, 3)
B. (1, -1) and (2, 2)
C. (1, 4) and (3, 2)
D. (2, -3) and (4, 5)
E. (3, -1) and (3, 1)

Question 28
Tim stands at the waterfront and holds a 30 cm ruler horizontally at eye level one metre in front of him. It lines up, so it appears to be exactly the same length as a cruise ship 1 km out to sea. How long is the cruise ship?

A.　299.7 m0　　　B.　300 m　　　C.　333 m　　　D.　29,970 m　　　E.　30,000 m

Question 29
Bob is twice as old as Kerry, and Kerry is three times as old as Bob's son. Their ages combined make 50 years. How old was Bob when his son was born?

A.　15　　　B.　20　　　C.　25　　　D.　30　　　E.　35

Question 30
In a healthy person, which one of the following has the highest blood pressure?

A. The vena cava
B. The systemic capillaries
C. The pulmonary artery
D. The pulmonary vein
E. The aorta
F. The coronary arteries

END OF SECTION

Section 1C

Passage 1-Birds

The following Passage is found in a book on nature published in 1899:

Five women out of every ten who walk the streets of Chicago and other Illinois cities, says a prominent journal, by wearing dead birds upon their hats proclaim themselves as lawbreakers. For the first time in the history of Illinois laws, it has been made an offence punishable by fine and imprisonment, or both, to have in possession any dead, harmless bird except game birds, which may be possessed in their proper season. The wearing of a tern, or a gull, a woodpecker, or a jay is an offence against the law's majesty, and any policeman with a mind rigidly bent upon enforcing the law could round up, without a written warrant, a wagon load of the offenders any hour in the day and carry them off to the lockup. What moral suasion cannot do, a crusade of this sort undoubtedly would.

Thanks to the personal influence of the Princess of Wales, the osprey plume, so long a feature of the uniforms of a number of the cavalry regiments of the British army, has been abolished. After Dec. 31, 1899, the osprey plume, by order of Field Marshal Lord Wolseley, is to be replaced by one of ostrich feathers. It was the wearing of these plumes by the officers of all the hussar and rifle regiments, as well as of the Royal Horse Artillery, which so sadly interfered with the crusade inaugurated by the Princess against the use of osprey plumes. The fact that these plumes, to be of any marketable value, have to be torn from the living bird during the nesting season induced the Queen, the Princess of Wales, and other ladies of the royal family to set their faces against the use of both the osprey plume and the aigrette as articles of fashionable wear.

Question 1

In 1899:

A. Women across the USA could be prosecuted for owning ornamental dead birds

B. There was a significant rise in female arrests in America

C. Possession of a dead gull could lead to trouble

D. Americans responded to law by citing the use of jays as ornamentation unfashionable

E. Delinquency across America increased

Question 2

Ostrich feathers were seen as preferable to osprey plumes because:

A. Ostriches are less intelligent birds

B. Ostriches are killed for their meat, so one might as well use their feathers

C. Queen Elizabeth has an especial love of ospreys

D. Harvesting osprey plumes was seen as an inhumane process

E. Ostrich feathers were of superior quality

Question 3

Which of the following is false, based on the passage?

A. Games birds could be possessed by citizens of Illinois all year round

B. Possessing a bird was not illegal in every circumstance

C. Wearing a woodpecker could lead to police action

D. Possessing certain birds could lead to a fine or imprisonment

E. All policemen would take action against any person wearing a prohibited bird

Question 4

Banning Osprey plumes in the UK's army was difficult because:

A. Many uniforms required them

B. The Princess did not have the authority to implement the ban

C. Her ultimate support was predominately female, and thus, their concerns seemed to have no relevance from the male domain of the army

D. It would be hard to differentiate between other regiments within the army, who were already wearing ostrich feathers

E. The production process of osprey plumes was easier

Question 5

Which of the following could NOT be legally owned in Illinois, according to the passage?

A. A live bird intended for personal ornamentation

B. A live bird intended for fighting

C. A dead bird of prey that had violently attacked you

D. Feathered garments

E. None of the above

Passage 2-Books

Gutenberg's father was a man of good family. Very likely the boy was taught to read. But the books from which he learned were not like ours; they were written by hand. A better name for them than books is 'manuscripts,' which means handwritings.

While Gutenberg was growing up, a new way of making books came into use, which was a great deal better than copying by hand. It was what is called block printing. The printer first cut a block of hardwood the size of the page that he was going to print. Then he cut out every word of the written page upon the smooth face of his block. This had to be very carefully done. When it was finished, the printer had to cut away the wood from the sides of every letter. This left the letters raised, as the letters are in books now printed for the blind. The block was now ready to be used. The letters were inked, the paper was laid upon them and pressed down. With blocks, the printer could make copies of a book a great deal faster than a man could write them by hand. But the making of the blocks took a long time, and each block would print only one page.

Gutenberg enjoyed reading the manuscripts and block books that his parents and their wealthy friends had, and he often said it was a pity that only rich people could own books. Finally, he determined to contrive an easy and quick way of printing.

Gutenberg, indeed, found this way and made the first movable-type printing press in Europe, with pieces from lead, tin, and antimony. Crucially. Gutenberg's innovation was confirmed by the production of the 'Gutenberg Bible'; the first major book printed with the movable-type printing press. With the invention came cheaper and higher quality books, thereby encouraging the development of printing presses across Europe, which in turn increased the number of books in supply. A new feature of the movable-type printing press was the advent of oil-based ink, which was more durable. This has been termed the 'Gutenberg Revolution'.

Question 6

Which of the following reasons can be inferred from the above passage to explain Gutenberg's desire to create a new way of printing?

A. It was a lucrative business to go into
B. He wanted to make text more accessible
C. He was tired of waiting for each book to be hand-written or block pressed and wanted quicker access to literature
D. He found the current books too costly for him to continue his reading habit
E. He wanted to spread his ideas across Europe

Question 7

Which of the following is **NOT** mentioned as a downside to block printing?

A. It exhausts the carver
B. It is intricate and demands attention to detail
C. It is a lengthy process
D. An individual block has limited utility
E. It requires the attention of an individual

Question 8

Which of the following statements is definitely true according to the above passage?

A. Gutenberg was taught to read as a boy
B. Gutenberg's father belonged to the aristocracy
C. Block printing was the predominant book manufacturing process whilst Gutenberg was growing up
D. Gutenberg's family was somewhat social
E. Gutenberg was deeply religious

Question 9

Which of the following is false, based on the passage?

A. Printing with the block process was a simple task of inking up the prepared block and pressing it down on a piece of paper, to make one page of the text
B. The movable-type printing press was the first of its kind in Europe
C. Oil based ink was not used in block printing
D. The ink lasted for longer in books made in the movable type printing press
E. More books were produced

Question 10

Which of the following statements are **NOT** supported by the above passage?

A. Manuscripts were beautifully crafted
B. 'Manuscripts' is an appropriate name for what it describes
C. Block printing is an appropriate name for what it describes
D. Having well off friends was a good way to expand your reading
E. It is probable that Gutenberg was educated

Passage 3- Norway

The following is taken from a book about Norway published in 1909:

'In a country like Norway, with its vast forests and waste moorlands, it is only natural to find a considerable variety of animals and birds. Some of these are peculiar to Scandinavia. Some, though only occasionally found in the British Isles, are not rare in Norway; whilst others (more especially among the birds) are equally common in both countries.

There was a time when the people of England lived in a state of fear and dread of the ravages of wolves and bears, and the Norwegians of the country districts even now have to guard their flocks and herds from these destroyers. Except in the forest tracts of the Far North, however, bears are not numerous, but in some parts, even in the South, they are sufficiently so to be a nuisance and are ruthlessly hunted down by the farmers. As far as wolves are concerned, civilisation is, fortunately, driving them farther afield each year and only in the most out-of-the-way parts are they ever encountered nowadays. Stories of packs of hungry wolves following in the wake of a sleigh are still told to the children in Norway, but they relate to bygone times—half a century or more ago, and such wild excitements no longer enter into the Norsemen's lives.

Yet, less ferocious animals give the people trouble enough, and amongst these may be mentioned the lynx and the wolverine, or glutton, each of which will make his supper off a sheep or a goat if he gets the chance. Of the two, the lynx is perhaps the worse poacher and his proverbial sharpness renders him difficult to catch. Not so the glutton, who, if he succeeds in crawling through a hole in the fence of a sheepfold, stuffs himself so full that he cannot get out again. I think that most of us would rather be called lynx-eyed than gluttonous, and certainly a lynx is a much handsomer beast than a glutton.

With the exception of the rabbit, all our English animals are found in Norway—the badger, fox, hare, otter, squirrel, hedgehog, polecat, stoat, and the rest of them. But besides these, there are little Arctic foxes and Arctic hares with bluish-grey coats in the summer and snowy-white ones in the winter. This change of colour is a provision of nature, rendering these particular animals, and some birds also, almost invisible among the snows. The ermine is another instance of this. In summer, he is just an ugly little brown stoat; but in winter, he comes out in pure white, with a jet-black tip to his tail, a skin worth a lot of money.'

Question 11
Which of the following is best supported by the above passage?

A. The variety of birds and animals to be found in Norway is unique to that country
B. The variety of birds and animals to be found in Norway is common to all European countries
C. By having forests, a country is more likely to have a variety of birds and animals
D. England and Norway have similar geographical features
E. All Norwegian animals can be found in England

Question 12
English people are described as:

A. Having been anxious about certain animals
B. Sceptical of bears
C. Living in fear of wolves
D. Developmentally behind the Norwegians
E. Worth a lot of money

Question 13
Bears are described as:

A.   Hunting   B.   Scavenging   C.   Damaging   D. Man-eating   E. 3Almost invisible

Question 14
Bears are also:

A. Numerous in all forest tracts
B. Numerous throughout the North
C. Numerous throughout the South
D. At risk in parts of Norway
E. In parts of England

Question 15
The Passage suggests:

A. The movement of wolves to the out-of-reach parts of Norway is beneficial
B. Wildlife currently threats Norwegian children
C. Regret at the loss of adventures
D. Norsemen particularly respect their natural surroundings
E. It would be preferable to be gluttonous

Question 16
Which of the following is an opinion?

A. It is better to be called lynx-eyed than gluttonous
B. The ermine changes colour
C. Norway has vast forests and waste moorlands
D. There are a variety of birds
E. English animals can be found in Norway

Passage 4- Colonialism

Most of the colonists who lived along the American seaboard in 1750 were the descendants of immigrants who had come in fully a century before; after the first settlements, there had been much less fresh immigration than many latter-day writers have assumed. According to Prescott F. Hall, "the population of New England ... at the date of the Revolutionary War ... was produced out of an immigration of about 20,000 persons who arrived before 1640," and we have Franklin's authority for the statement that the total population of the colonies in 1751, then about 1,000,000, had been produced from an original immigration of less than 80,000.

Even at that early day, indeed, the colonists had begun to feel that they were distinctly separated, in culture and customs, from the mother-country and there were signs of the rise of a new native aristocracy, entirely distinct from the older aristocracy of the royal governors' courts. The enormous difficulties of communication with England helped to foster this sense of separation.

The round trip across the ocean occupied the better part of a year, and was hazardous and expensive; a colonist who had made it was a marked man—as Hawthorne said, "the petit maître of the colonies." Nor was there any very extensive exchange of ideas, for though most of the books read in the colonies came from England. The great majority of the colonists, down to the middle of the century, seem to have read little save the Bible and biblical commentaries, and in the native literature of the time, one seldom comes upon any reference to the English authors who were glorifying the period of the Restoration and the reign of Anne. "No allusion to Shakespeare," says Bliss Perry, "has been discovered in the colonial literature of the seventeenth century, and scarcely an allusion to the Puritan poet Milton." Benjamin Franklin's brother, James, had a copy of Shakespeare at the *New England Courant* office in Boston, but Benjamin himself seems to have made little use of it, for there is not a single quotation from or mention of the bard in all his voluminous works. "The Harvard College Library in 1723," says Perry, "had nothing of Addison, Steele, Bolingbroke, Dryden, Pope, and Swift, and had only recently obtained copies of Milton and Shakespeare....Franklin reprinted 'Pamela' and his Library Company of Philadelphia had two copies of 'Paradise Lost' for circulation in 1741, but there had been no copy of that work in the great library of Cotton Mather."

Question 17

Which of the following is true according to the passage?

A. Over half of the 1750 colonists that lived on the American seaboard had genetic links to immigrants who had arrived a century ago

B. Most of the books on board ships were Bibles and biblical commentaries

C. The colonists had poor communication skills

D. The colonists disliked the English

E. Many colonists visited England after immigrating to America

Question 18

Which of the following statements is supported by the above passage?

A. According to Hall, America's population at the date of the Revolutionary War could be entirely traced back to 20,000 immigrants

B. The population in the 1751 colonies was over ten times the original immigration that moved there

C. According to Hall, in 1751, the population in the American colonies was one million

D. According to Hall, 80,000 people led to a population of 1,000,000

E. Most of the population on the American seaboard immigrated there

Question 19

According to the passage, the new aristocracy that existed in the colonies was:

A. Similar to the England's

B. Similar to European aristocratic systems in general

C. Not based in royal governors' courts

D. Not based on genetic lines

E. Based on custom

Question 20

Which of these is **NOT** given as a reason for the sense of separation from England?

A. Travel between America and England was costly

B. The English saw the early colonists as backwards

C. Travel between America and England was slow

D. Travel between America and England was dangerous

E. Lack of an exchange of ideas

Question 21

What did the author mean by the word 'allusion' in the passage?

A. An implication

B. An illusion

C. Deference

D. Thoughts

E. Reference

Passage 5 – Rorschach

When discussing his famous character Rorschach, the antihero of 'Watchmen', Moore explains, 'I originally intended Rorschach to be a warning about the possible outcome of vigilante thinking. But an awful lot of comic readers felt his remorseless, frightening, psychotic toughness was his most appealing characteristic – not quite what I was going for.' Moore misunderstands his own hero's appeal within this quotation: it is not that Rorschach is willing to break little fingers to extract information, or that he is happy to use violence, that makes him laudable. The Comedian, another 'superhero' within the alternative world of Watchmen, is a thug who has won no great fan base; his remorselessness (killing a pregnant Vietnamese woman), frightening (attempt at rape), psychotic toughness (one only has to look at the panels of him shooting out into a crowd to witness this) is repulsive, not winning. This is because The Comedian has no purpose: he is a nihilist, and as a nihilist, denies any potential meaning to his fellow man, and so to the comic's reader. Everything to him is a 'joke', including his self, and consequently, his own death could be seen as just another gag.

Rorschach, on the other hand, does believe in something: he questions if his fight for justice 'is futile?' then instantly corrects himself, stating 'there is good and evil, and evil must be punished. Even in the face of Armageddon I shall not compromise in this.' Jacob Held, in his essay comparing Rorschach's motivation with Kantian ethics, put forward the postulation that 'perhaps our dignity is found in acting as if the world were just, even when it is clearly not.' Rorschach then causes pain in others not because he is a sadist, but because he feels the need to punish wrong and to uphold the good, and though he cannot make the world just, he can act according to his sense of justice - through the use of violence.

Question 22

Which of the following best describes 'Watchmen'?

A. A book that contains only vicious characters
B. An expression of despair when contemplating an imperfect world
C. An example of how an author's intentions are not always realised
D. A book that accidentally glamorises violence
E. A book which highlights the ideal hero

Question 23

Which of the following best accords with the view of the author?

A. All heroes use a minimum of violence
B. No hero uses violence
C. Heroes aim to uphold good
D. A hero should have a sense of purpose
E. It is impossible to be both a comedian and a hero

Question 24

Which of the following best articulates the view put forward by Jacob Held?

A. We find dignity through just actions
B. If one decides to behave as though the world is fair, this may lead to a discovery of self-worth
C. It is shameful to view the world as corrupt
D. Self-value can only be found in madness
E. All means should be used in the pursuit of justice, including violence

END OF SECTION

Section 2

YOU MUST ANSWER <u>ONLY</u> <u>ONE</u> OF THE FOLLOWING QUESTIONS

Question 1

"The eternal mystery of this world is its comprehensibility"

To what extent is the world comprehensible?

Question 2

"The greatest obstacle to learning is education"

Argue for or against this statement.

Question 3

"Strive not to be a success, but to be of value"

To what extent is it possible to be "a success", but to have little value?

Question 4

"Why tell the truth if a lie is better for all concerned?"

In what circumstances can dishonesty be justified?

END OF PAPER

Mock Paper B

Section 1A

Question 1

Joseph changes jobs and gets a basic pay cut of 5%, but his tax-free monthly bonus increases from £40 to £90. He also changes tax bracket, so instead of his 10% flat rate, he pays 20% tax on all income over £10,000 pa. Joseph's current weekly take-home pay is £560 per week, exclusive of bonus.
What will his new annual take-home pay be, to the nearest hundred pounds?

A.   £25,800   B.   £26,500   C.   £27,000   D.   £29,300   E.   £31,300

Question 2

Peter books a return flight to Dubai for £725. The flight is refundable, but there is a fee of £45 payable for cancelling. Peter notices as time passes, the remaining tickets on the same plane are becoming cheaper. He decides to cancel his flight, booking a new one for £530 through the same provider. Once again, he sees prices have fallen, so he cancels this flight but can only buy a new one for £495.
What is his overall saving, relative to the original price paid?

A.   £110   B.   £140   C.   £150   D.   £195   E.   £230

Question 3

You have three bags, each containing four balls numbered with single digit numbers. Bag A contains even numbers only, Bag B contains odd numbers only, and Bag C contains the numbers 2, 5, 6 and 8. You take a ball from Bag B and put it into Bag C; then you then take a ball from Bag C and put it into Bag A. You draw a ball at random from Bag A
What is the probability that this ball is an odd number?

A. $^1/_{25}$ B. $^2/_{25}$ C. $^3/_{25}$ D. $^4/_{25}$ E. $^1/_5$

Question 4

The price of bread rises by 40% due to a poor grain harvest. This is later reduced by 20% due to a government farming subsidy. Dave buys three loaves of bread and gets a fourth free because of a discount in the shop. How much did he pay per loaf of bread? Express your answer as a percentage of the original price.

A.   66%   B.   84%   C.   92%   D.   98%   E.   110%

Question 5

Sam notes that the time on a normal analogue clock is 2120hrs. What is the smaller angle between the hands on the clock?

A.   130°   B.   140°   C.   150°   D.   160°   E.   170°

Question 6

Sam needs to measure out exactly 4 litres of water into a tank. He has two pieces of equipment – a bucket that holds 5 litres and a one that holds 3 litres, with no intermediate markings. Is it possible to measure out 4 litres? If so, how much water is needed in total in order to measure the 4 litres?

A.   4 litres   B.   7 litres   C.   8 litres   D.   10 litres   E.   Not possible with this equipment

Question 7

A librarian is sorting books into their correct locations. All history books belong to the right of all science books. Science books are divided into five locations: engineering, biology, chemistry, physics and mathematics (in order from right to left). Art books are located between engineering and sport, and sport books between art and history. Literature books are to the right of art books.

What can be certainly said about the location of literature books?

A. They are located between art and history books
B. They are located to the left of history books
C. They are located between mathematics and art
D. They are located to the right of engineering
E. They are not located to the left of sport

Question 8

Many people choose not to buy brand new cars, as buying brand new has significant disadvantages. Most importantly, a car's value drops substantially at the moment it is first driven on the road. Even though a car is virtually unchanged by these first few miles, the potential resale value is significantly reduced. Therefore, it is better to buy second hand cars, as their value does not drop so much immediately after purchase

Which of the following best represents the main conclusion of this passage?

A. There are many equal reasons to avoid buying brand new cars
B. Cars that have driven lots of miles should be avoided
C. The rapid loss of value in new cars makes buying second-hand a wise choice
D. Second hand cars are at least as good as new ones
E. New cars should not be driven to ensure they keep their resale value

Question 9

James is a wine dealer specialising in French wine. From his original stock of 2,000 bottles in one cellar, he sells 10% to one customer and 20% of the remaining wine to another customer. He makes £11,200 profit from the two transactions combined. What is the average profit per bottle?

A.   £18   B.   £20   C.   £22   D.   £24   E.   £26

Question 10

Why should we bother exploring deep into the oceans? The programmes are very expensive, and seldom produce any results which benefit normal people. Instead, we should invest resources into supporting people in trouble, rather than wasting money on needless exploration.

Which of the following, if true, would most weaken the above argument?

A. Ocean exploration is less expensive than space exploration, which people are generally happy with
B. Ocean exploration provides fascinating information about bizarre life forms
C. Exploration has led to the discovery of new chemicals which have been used for many new medically useful drugs
D. Exploring into the oceans is safe, given modern submarine technology
E. Money is useful to help people in trouble

Question 11

Many good quality pieces of old furniture are considered 'timeless' – they are used and enjoyed by many people today, and this is expected to continue for many generations to come. However, most of this furniture dates back to previous eras, and modern furniture does not fall under the 'timeless' category of being enjoyed for many years to come.
Which of the following is the main flaw in the argument?

A. There may be many factors which make furniture good
B. There used to be more furniture makers than today
C. No evidence is given to tell us old furniture is better than new
D. Old furniture is desirable for other reasons than its quality
E. We cannot yet tell whether new furniture will become 'timeless'

Question 12

Red wine is thought to be much healthier than beer because it contains many antioxidants, which have been shown to be beneficial to health. Many red wines are produced in Southern France and Italy; therefore, it is no surprise that residents there have a greater life expectancy than in the UK and Germany, which are predominantly beer producing countries
Which of the following is an assumption of the above argument?

A. Italian people drink red wine
B. Antioxidants are beneficial for health
C. British people prefer beer to red wine
D. Beer is not produced in Italy
E. Italian life expectancy is greater than in the UK

Question 13

Hannah, Jane and Tom are travelling to London to see a musical. Hannah catches the train at 1430. Jane leaves at the same time as Hannah, but catches a bus which takes 40% longer then Hannah's train. Tom also takes a train, and the journey time is 10 minutes less then Hannah's journey, but he leaves 45 minutes after Jane leaves. He arrives in London at 1620. At what time will Jane arrive in London?

A. 1545   B. 1600   C. 1615   D. 1700   E. 1715

Question 14

At a show, there are two different ticket prices for different seats. The cost is £10 for a standard seat, and £16 for a premium view seat. The total revenue from a show is £6,600, and the total attendance was 600. How many premium view seats were purchased?

A.   60   B.   100   C.   140   D.   180   E.   240

Question 15

The moon orbits the Earth once every 28 days. Between 20th January and 23rd April inclusive, how many degrees has the Moon turned through? This is not a leap year.

A.   1010   B.   1100   C.   1210   D.   1500   E.   1620

Question 16

Drama academies are special schools students can go to in order to learn performing arts. These schools are only available to the most skilled young performers, and aim to give students the best training in the arts, whilst still covering mainstream academic subjects. However, many parents are reluctant for their children to attend such academies, as they feel the academic teaching will be worse than a standard school.

Which of the following, if true, would most weaken the above argument?

A. Most top actors attended a drama academy as children
B. There is as much time dedicated to academic work in drama academies as there is in normal schools
C. The academic work comprises a greater proportion of the study time than drama related activities
D. Most children are keen to attend a drama academy if given the opportunity
E. 80% of students at drama academies attain higher than average GCSE scores

Question 17

Anil and Suresh both leave point A at the same time. Anil travels 5km East then 10km North. Anil then travels a further 1km North before heading 3km West. Suresh travels East for 2km less than Anil's total journey distance. He then heads 13km North, before pausing and travelling back 2km South. How far, as the crow flies, are the two men now apart?

A.   11   B.   12   C.   13   D.   15   E.   17

Question 18

Building foundations are covered by 14cm of concrete. A builder thinks this is too thick, and grinds down the concrete by an amount three times the thickness of the concrete which he eventually leaves. What is the remaining thickness of concrete?

A.   1.5 cm   B.   2.0 cm   C.   2.5 cm   D.   3.0 cm   E.   3.5 cm

Question 19

Chris leaves his house to go and visit Laura, who lives 3 miles away. He leaves at 1730 and walks at 4mph towards Laura's house, stopping for one 5-minute to chat to a friend. Meanwhile Sarah also wants to visit Laura. She sets off from her house 6 miles away at 1810, driving in her car and averaging a speed of 24mph. Who reaches the house first and with how long do they wait for the other person?

A. Chris, and waits 5 mins for Sarah
B. Chris, and waits 10 mins for Sarah
C. Sarah, and waits 5 mins for Chris
D. Sarah, and waits 10 mins for Chris
E. They both arrive at the same time

Question 20

Illegal film and music downloads have increased greatly in recent years. This causes significant harm to the relevant industries. Many people justify this to themselves by telling themselves they are only diverting money away from wealthy and successful singers and actors, who do not need any more money anyway. But in reality, illegal downloads are deeply harming the music industry, making many studio workers redundant and making it difficult for less famous performers to make a living.

Which of the following best summarises the conclusion of this argument?

A. Unemployment is a problem in the music industry
B. Taking profits away from successful musicians does more harm than good
C. Studio workers are most affected by illegal downloads
D. Illegal downloads cause more harm than people often think
E. Buying music legally helps keep the music industry productive

Question 21

40,000 litres of water will extinguish two typical house fires. 70,000 litres of water will extinguish two house fires and three garden fires. There is no surplus water
Which statement is **NOT** true?

A. A garden fire can be extinguished with 12,000 litres, with water to spare.
B. 20,000 litres are sufficient to extinguish a normal house fire.
C. A garden fire requires only half as much water to extinguish as a house fire.
D. Two houses and four garden fires will need 80,000 litres to extinguish.
E. Three houses and ten garden fires will need 140,000 litres to extinguish.

Question 22

A car travels at $20ms^{-1}$ for 30 seconds. It then accelerates at a constant rate of $2ms^{-2}$ for 5 seconds, then proceeds at the new speed for 20 seconds before braking with constant deceleration of $3ms^{-2}$ to a stop. What distance is covered in total?

A.   1325 m   B.   1350 m   C.   1375 m   D.   1425 m   E.   1475 m

END OF SECTION

Section 1B

Question 1
Which of the following statements are true?

1. Natural selection always favours organisms that are faster or stronger.
2. Genetic variation creates different adaptations to the environment.
3. Variation is purely due to genetics.

A.   Only 1   B.   Only 2   C.   Only 3   D.   1 and 2   E.   2 and 3

F.   1 and 3   G.   All of the above   H.   None of the above

The diagram to the right is necessary for questions 2 – 3:

Question 2
Which of the following numbers indicate where amylase functions?
A. 1 only
B. 3 only
C. 1 and 3
D. 1 and 5
E. 2 and 4
F. 3 and 4
G. 5 and 6

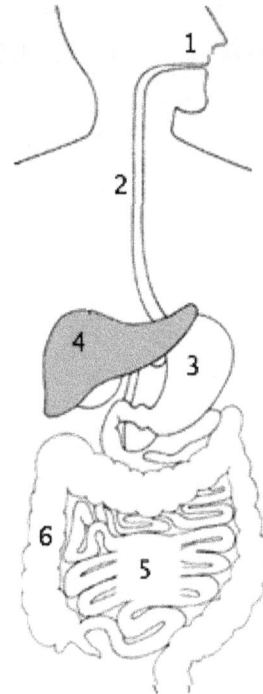

Question 3
In which of the following does the majority of chemical digestion occur?
A. 1
B. 2
C. 3
D. 4
E. 5
F. 6
G. None of the above

Question 4
Which of the following statements is false?

A. A nuclear power plant may have an accident if free neutrons in a fuel rod aren't captured.
B. Humans cannot currently harness the energy from nuclear fusion.
C. Uncontrolled nuclear fission leads to a large explosion.
D. Mass is conserved during nuclear explosions caused by nuclear bombs.
E. Nuclear fusion produces much more energy than nuclear fission.

Question 5

Make m the subject: $T = 4\pi\sqrt{\dfrac{(M+3m)l}{3(M+2m)g}}$

A. $m = \dfrac{16\pi^2 lM - 3gMT^2}{48\pi^2 l - 6gT^2}$

B. $m = \dfrac{16\pi^2 lM - 3gMT^2}{6gT^2 - 48\pi^2 l}$

C. $m = \dfrac{3gMT^2 - 16\pi^2 lM}{6gT^2 - 48\pi^2 l}$

D. $m = \dfrac{4\pi^2 lM - 3gMT^2}{6gT^2 - 16\pi^2 l}$

E. $m = \left(\dfrac{16\pi^2 lM - 3gMT^2}{6gT^2 - 48\pi^2 l}\right)^2$

The following information applies to questions 6 – 7:

The diagram below shows the genetic inheritance of colour-blindness, which is inherited in a sex-linked recessive manner [transmitted on the X chromosome and requires the absence of normal X chromosomes to result in disease]. X^B is the normal allele and X^b is the colour-blind allele.

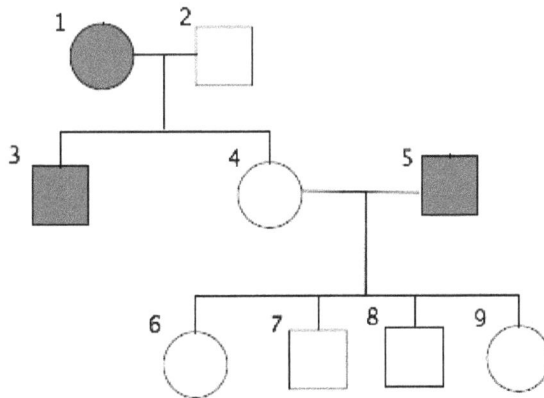

Question 6
What is the genotype of the individual marked 4?

A. $X^B X^b$ B. $X^B X^B$ C. $X^b X^b$ D. $X^B Y$ E. $X^b Y$ F.   None of the above

Question 7
If 8 were to reproduce with 6, what is the probability of producing a colour-blind boy?

A.   100%   B.   75%   C.   50%   D.   25%   E.   12.5%   F.   0%

Question 8
The mean of a set of 11 numbers is 6. Two numbers are removed and the mean is now 5. Which of the following is not a possible combination of removed numbers?

A.   1 and 20   B.   6 and 9   C.   10 and 11   D.   15 and 6   E.   19 and 2

Question 9

Find the values of angles b and c.

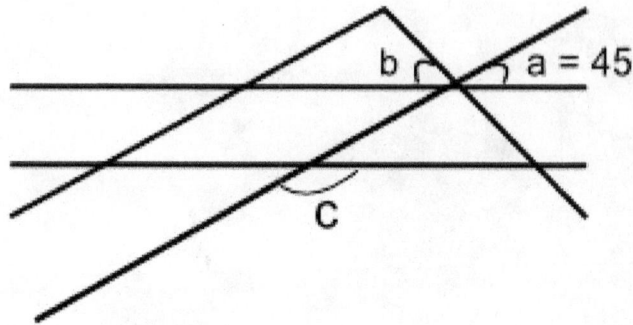

A.   45° and 135°   B.   45° and 130°   C.   50° and 135°   D.   55° and 130°   E.   More   information

Question 10

Evaluate: $\frac{3.4 \times 10^{11} + 3.4 \times 10^{10}}{6.8 \times 10^{12}}$

A. 5.5×10^{-12} B. 5.5×10^{-2} C. 5.5×10^{1} D. 5.5×10^{2} E. 5.5×10^{10} F. 5.5×10^{12}

The following information applies to questions 11 – 12:

In pea plants, colour and stem length are inherited in an autosomal manner. The allele for yellow colour, Y, is dominant to the allele for green colour, y. Furthermore, the allele for tall stem length, T, is dominant to short stem length, t.

When a pea plant of unknown genotype is crossed with a green short-stemmed pea plant, the progeny are 25% yellow + tall-stemmed plants, 25% yellow + short-stemmed plants, 25% green + tall-stemmed plants and 25% green + short-stemmed plants.

Question 11

What is the genotype of the unknown pea plant?

A.   Yytt   B.   YyTt   C.   YyTT   D.   yyTt   E.   yyTT   F.   yytt
G.   More information needed

Question 12

Taking both colour and height into account, how many different combinations of genotypes and phenotypes are possible?

A.   6 genotypes and 3 phenotypes   B. 8 genotypes and 3 phenotypes   C. 8 genotypes and 4 phenotypes
D.   9 genotypes and 4 phenotypes   E. 9 genotypes and 3 phenotypes   F. 10 genotypes and 3 phenotypes

Question 13

Evaluate the following expression: $\left(\frac{6}{8} \times \frac{7}{3} \div \frac{7}{5} \times \frac{2}{6}\right) \times 0.40 \times 15\% \times 5\% \times \pi \times \left(\sqrt{e^2}\right) \times 0.20 \times (e\pi)^{-1}$

A. $\frac{4}{55}$ B. $\frac{8}{770}$ C. $\frac{9}{4,000}$ D. $\frac{8}{54,321}$ E. $\frac{9}{67,800}$

Question 14

What is the **MOST** important reason for each cell in the human body to have an adequate blood supply?

A. To allow protein synthesis.
B. To receive essential minerals and vitamins for life.
C. To kill invading bacteria.
D. To allow aerobic respiration to take place.
E. To maintain an optimum cellular temperature.
F. To maintain an optimum cellular pH.

Question 15

A man drives along a road as shown in the figure below.

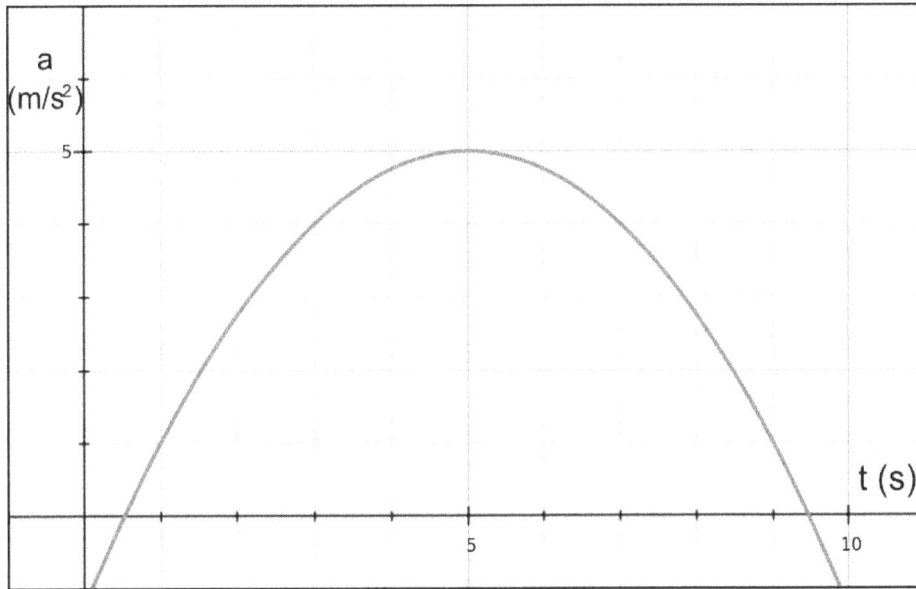

Which of the following statements is true?

A. He drives a total of 30 m.
B. He has an average velocity of 30 m/s.
C. He has a final velocity of 30 m/s.
D. He has an average acceleration of 30 m/s^2.
E. His velocity decreases between 5 and 9 seconds.

Question 16

A tangent line to a circle of radius 3 metres intersects with another line 4 metres from its tangent point. How far is this point of intersection from the centre of the circle?

A. 1 metres
B. 3 metres
C. 5 metres
D. 7 metres
E. 9 metres

Question 17

In relation to the human genome, which of the following are correct?

1. The DNA genome is coded by 4 different bases.
2. The sugar backbone of the DNA strand is formed of glucose.
3. DNA is found in the nucleus of bacteria.

A. 1 only   B. 2 only   C. 3 only   D. 1 and25   E. 1 and 2

F. 2 and 3   G. 1,2 and 3

Question 18

Animal cells contain organelles that take part in vital processes. Which of the following is true?

1. The majority of energy production by animal cells occurs in the mitochondria.
2. The cell wall protects the animal cell membrane from outside pressure differences.
3. The endoplasmic reticulum plays a role in protein synthesis.

A. 1 only   B. 2 only   C. 3 only   D. 1 and 2   E. 2 and 3

F. 1 and 3   G. 1,2 and 3

Question 19

With regards to animal mitochondria, which of the following is correct?

A. Mitochondria are not necessary for aerobic respiration.
B. Mitochondria are the sole cause of sperm cell movement.
C. The majority of DNA replication happens inside mitochondria.
D. Mitochondria are more abundant in fat cells than in skeletal muscle.
E. The majority of protein synthesis occurs in mitochondria.
F. Mitochondria are enveloped by a double membrane.

Question 20

In relation to bacteria, which of the following is **FALSE**?

A. Bacteria always lead to disease.
B. Bacteria contain plasmid DNA.
C. Bacteria do not contain mitochondria.
D. Bacteria have a cell wall and a plasma membrane.
E. Some bacteria are susceptible to antibiotics.

Question 21

In relation to bacterial replication, which of the following is correct?

A. Bacteria undergo sexual reproduction.
B. Bacteria have a nucleus.
C. Bacteria carry genetic information on circular plasmids.
D. Bacterial genomes are formed of RNA instead of DNA.
E. Bacteria require gametes to replicate.

Question 22

Which of the following are correct regarding active transport?

A. ATP is necessary and sufficient for active transport.
B. ATP is not necessary but sufficient for active transport.
C. The relative concentrations of the material being transported have little impact on the rate of active transport.
D. Transport proteins are necessary and sufficient for active transport.
E. Active transport relies on transport proteins that are powered by an electrochemical gradient.

Question 23

Concerning mammalian reproduction, which of the following is **FALSE**?

A. Fertilisation involves the fusion of two gametes.
B. Reproduction is sexual, and the offspring display genetic variation.
C. Reproduction relies upon the exchange of genetic material.
D. Mammalian gametes are diploid cells produced via meiosis.
E. Embryonic growth requires carefully controlled mitosis.

Question 24

Robert has a box of building blocks. The box contains 8 yellow blocks and 12 red blocks. He picks three blocks from the box and stacks them up high. Calculate the probability that he stacks two red building blocks and one yellow building block, in **any** order.

A. $\frac{8}{20}$ B. $\frac{44}{95}$ C. $\frac{11}{18}$ D. $\frac{8}{19}5$ E. $\frac{12}{20}$ F. $\frac{35}{60}$

Question 25

Solve $\frac{3x+5}{5} + \frac{2x-2}{3} = 18$

A. 12.11   B. 13.49   C. 13.95   D. 14.2   E. 19   F. 265

Question 26

Solve $3x^2 + 11x - 20 = 0$

A. 0.75 and $-\frac{4}{3}$ B. -0.75 and $\frac{4}{3}$ C. -5 and $\frac{4}{3}$ D. 5 and $\frac{4}{3}$ E. 12 only   F. -12 only

Question 27

Express $\frac{5}{x+2} + \frac{3}{x-4}$ as a single fraction.

A. $\frac{15x-120}{(x+2)(x-4)}$ B. $\frac{8x-26}{(x+2)(x-4)}$ C. $\frac{8x-14}{(x+2)(x-4)}$ D. $\frac{15}{8x}$ E. 24   F. $\frac{8x-14}{x^2-8}$

Question 28

The value of p is directly proportional to the cube root of q. When p = 12, q = 27. Find the value of q when p = 24.

A. 32   B. 64   C. 124   D. 128   E. 216   F. 1728

Question 29
Write 72^2 as a product of its prime factors.

A. $2^6 \times 3^4$ B. $2^6 \times 3^5$ C. $2^4 \times 3^4$ D. 2×3^3 E. $2^6 \times 3$ F. $2^3 \times 3^2$

Question 30
Calculate: $\dfrac{2.302 \times 10^5 + 2.302 \times 10^2}{1.151 \times 10^{10}}$

A. 0.0000202   B. 0.00020002   C. 0.00002002   D. 0.00000002   E. 0.000002002   F. 0.000002002

END OF SECTION

Section 1C

Passage 1 – Gambler's Fallacy

The gambler's fallacy is a logical fallacy, where an independent event becomes more predictable the more it is repeated and, of course, takes place in the context of gambling. It can be demonstrated with the example of a dice being thrown. On the first throw yielding a score of 5, the second yielding a 5, what is the probability of another 5 coming up? Some may think it nearly impossible. Indeed, the odds are low when considering all 3 throws together but when considering just that third throw, the odds are still 1 out of 6, just as it was for the first and second throws. Crucially, each throw is independent of each other, so a previous throw has no bearing on the next one. Therefore, contrary to instinct, the independent event becomes no less independent on a second try. However, this is a frequently made mistake by gamblers.

Another name for the gambler's fallacy is the Monte Carlo fallacy, which is named after a famous example of it occurring at the Monte Carlo Casino in 1913. The night proceeded as a normal one until it was noticed that the roulette ball had fallen on black for a number of rounds. As it kept on coming on black, more and more gamblers were putting their money in - surely the ball would fall on red soon? It didn't. 26 spins in a row, in fact, fell on black, leading to losses in the millions. Gamblers believed that the odds were stacked in their favour, even after a few black runs. However, as a mathematical probability, it makes no difference how many blacks or reds there are: at each round, there's always a 50% chance of it being either black or white.

The basic laws of probability are taught at GCSE Mathematics level, so it should not be a problem for those who paid attention at school, which raises a question as to why it remains such a prevalent issue. Indeed, research has shown that sub-conscious processes may have a role to play. In the Journal of Experimental Psychology, researchers conducted a study of participants on the gambler's fallacy. The participants were split into two groups: the experimental group and the control group. The gambler's fallacy was explicitly explained to the experimental group and they were told not to rely on previous runs when making guesses. The control group was told nothing about this, though. The researchers then showed the participants a re-shuffled deck of cards and asked them to guess which shape would come next in the sequence. The results showed that the responses did not differ between the different groups, thereby indicating that the experimental group was not swayed by knowledge of the gambler's fallacy.

In particular, though, Roney and Trick showed in 2003 that grouping events, such that the next event appears as if it comes at the beginning of the next sequence, may overcome the gambler's fallacy. In the study, participants were shown a sequence of 6-coin tosses, with the final 3 flips being heads. One group were asked to predict the seventh flip. The other group was asked to predict the first event for the next sequence. Those in the former group tended to predict tails more. Accordingly, viewing independent events as a 'beginning' rather than as part of a sequence can help abate the gambler's fallacy.

Question 1
Which of the following best explains the word 'fallacy'?

A. A gambling addiction
B. A belief
C. A mistaken belief
D. When a dice falls independently of the previous dice
E. A lie

Question 2
Which of the following is the best definition for the Monte Carlo fallacy?

A. A gambling addiction
B. The events in Monte Carlo in 1913
C. The view that a previous event influences a future event
D. The view that a future event is thrown into doubt by a previous event
E. The circumstances around an independent event

Question 3
Why did gamblers place money on the red in the Monte Carlo Casino incident in 1913?

A. They knew they were going to get a windfall
B. The red was more popular than the black in those times
C. 26 spins in a row were on black
D. They thought that the fact of the ball falling on black so many times increased their odds of winning on red
E. They disagreed with the gambler's fallacy

Question 4
Which of the following can be assumed about the research in the Journal of Experimental Psychology?

A. The experimental group did not listen to what the researchers said to them
B. The participants represented all areas of society
C. The gambler's fallacy was tested
D. Participants were shown at least one card
E. The researchers were biased

Question 5
Which of the following can be inferred from the third paragraph?

A. The gambler's fallacy cannot be solved
B. The gambler's fallacy was proved to exist by that research
C. The research took place after 1913
D. Teaching probability does not necessarily alleviate the gambler's fallacy
E. Researchers conducted a study of participants

Question 6
In the final paragraph, why did those in the first group tend to predict tails more?

A. There was a head in the time before
B. They did not believe in the gambler's fallacy
C. They believed in the gambler's fallacy
D. The gambler's fallacy
E. A matter of probability

Passage 2 – Human Organ Sales

In the year 2013/14, over 4,000 organ transplants took place, 1,146 of which consisted of donations from living individuals. Despite recent year-on-year increases, the current rate of organ transplants is not enough to satisfy demand for them, and this has severe consequences: approximately three people die in the UK every day from not having an organ transplant.

NHS Blood and Transplant suggest that there needs to be a revolution in social attitudes towards donation, but such a revolution needs a radical change in tactics. A public campaign to increase awareness of joining the organ donation register has taken heed. Indeed, any member of the public can register to donate their organs in the event of death. However, take up has been relatively low. In order to help counter the shortage, the Welsh government have adopted a system of presumed consent such that a person is deemed to consent to their organs being donated unless they have indicated otherwise (in which case, their organs will not be donated). Again, though, whether such a scheme will alleviate the shortage is unclear.

Accordingly, it has been proposed that a marketplace for organs be developed, where people are allowed to provide organs and be paid for them. This would acknowledge that the shortage is a major public health problem and encourage donors to come forward. In response to such suggestions in the past, opponents have maintained that it is immoral, but they tend only to involve either no or superficial consideration of the countervailing policy considerations and critically do not address the shortage of organs. Indeed, one can actually query whether it is moral to allow people to die in circumstances in which death can be avoided.

The positive case for allowing the sale of organs is generally based on the existence of a shortage, but there are other useful considerations too. The libertarian argument is based on the view that adults exercising their free will should be able to decide what to do with their own bodies. Accordingly, should they wish to give their organs in return for a price, there should not be an objection to it unless there are strong reasons against it. Freedom and respect for personal autonomy demand no less.

A regular assertion is that allowing the sale of organs will exploit the poor; they would be in the most need to cash in on their spare organs which could be at the expense of their health. However, they can still give their organs at present via a donation – why should there be a difference in the health risks between getting money for giving your organ and getting nothing for it? If the current risk framework for living transplants was used to determine whether one can sell their organs, giving an organ for money will not be more dangerous and riskier. In fact, providing money for organs is fairer than the current system – the providers of organs are compensated for the organ, the health risks, and their time. The involvement of money should not make it immoral – the surgeons carrying out the transplant, the transportation officers, and nurses are all paid for carrying out a transplant, but no one would argue they're immoral.

Far from being immoral, the individual providing the organ would be saving someone else's life. Providing economic compensation should not make a difference. While opponents suggest that the poor would be exploited, it is incumbent on them to elaborate on how this is a different form of exploitation to working in high-risk occupations such as deep-sea diving, in the military or in a coal mine. The health risks of these occupations far outweigh the risk undertaking an operation to remove an organ, such as a kidney. Far from being exploited, people, however rich they are, should be able to give, should they want to.

Of course, this is not to say that any market in organs should be unregulated. In fact, quite the contrary is suggested by proponents of this. A heavily regulated market and safeguards to ensure fully informed consent and protection for the vulnerable are vital. Moreover, the system as proposed by Erin and Harris (2003) would fit well with the current system. There would only be one buyer of organs, the National Health Service, and organs could be distributed in accordance with existing priority rules, based on need and fairness.

This does not make the medical profession less caring or infringe on morality, it is, in fact, our moral imperative to ensure that we protect as many lives as possible. On a consequentialist perspective, the benefits of a market for human organs are clear but the drawbacks are not.

Question 7

In which of the following circumstances would the libertarian argument not support the sale of organs?

A. Where it is not provable that organ transplants would save lives
B. Where it is against nature
C. For those who do not give informed consent
D. Where there is a risk to the donor
E. When there is not a shortage

Question 8

Which of the following is true?

A. The Welsh government impose organ donation on individuals after their death
B. Organs that are sold will go to the highest bidder
C. Organ donation is moral
D. A person can refuse to have their organs donated after death in Wales
E. A marketplace for organs would be unethical

Question 9

Which of the following is an argument in the passage?

A. People should be able to sell their organs
B. Opponents have maintained that it is immoral
C. A market for organs should not exist
D. The libertarian argument should prevail over all others
E. The health risk of organ donation is outweighed by the benefit to the recipient

Question 10

Which of the following is an opinion?

A. The poor would be exploited
B. In 2013/14, over 4,000 organ transplants took place
C. There is a shortage of organs
D. There are not enough donors to satisfy demand
E. A public campaign has taken place to encourage organ donations

Question 11

Which of the following is not advanced as an argument, implicit or explicit, by the author?

A. Poorer individuals should be allowed to donate their organs should they wish
B. There is little difference between selling an organ and working in a high-risk job
C. A human organ market would be fair
D. It would be better for the poor as opposed to the rich, to donate their organs
E. The health risks are not increased for an organ transplant which is sold

Passage 3 – Free Speech

If someone says something you don't like, what do you do? Stop them from speaking or let them speak, but state your disagreement with them? In day-to-day conversations, either option does not appear to make much difference but on a larger scale, the repercussions for freedom of expression are enormous.

Either course of action has different consequences and implications for society at large. Take the former option and the other person won't be heard – in the event of the latter option, that person may be heard, and the public may disagree with it but at least the public would be able to make up their own minds on the matter. Were the public to agree with the position, surely that is a right people have, regardless of your opinion.

Student unions in British Universities have been particularly vociferous in clamping down on speakers who might offend students. Indeed, many question the extent to which a person has a right to offend another. However, as Louise Richardson has pointed out, education is not comfortable. On the contrary, it is about tackling ideas and arguments you don't like, "confronting the person you disagree with and trying to change their mind. This isn't a comfortable experience, but it is a very educational one."

Student unions are in a position of significant power in university circles, in being able to organise protests and decide on who speaks in their debates. A number of student unions operate a 'no-platform' policy, where an individual is prohibited from speaking at events. This may be due to the individual holding extremist views, but someone has to make a judgement as to what counts as extremist. This can potentially be a serious infringement of a person's freedom of speech if misused.

Indeed, student union leaders have taken contentious actions recently. For example, controversial writer, Germaine Greer, gave a speech at Cardiff University. While this was accompanied with a protest, the university's student union also started a petition to ban the writer from speaking in the first place. Most individuals may well disagree with Greer's arguments, which indeed are extreme.

What was more surprising, however, was that Peter Tatchell, a gay rights campaigner, was declared transphobic and racist by the National Union of Student's LGBT (Lesbian, Gay, Bisexual and Transgender) officer, who refused to share a platform with him. The officer justified her statement because Tatchell signed an open letter denouncing the NUS' no platform policy and supporting the free speech of individuals such as Germaine Greer. This shows a surprising mind-set in the student's union. If mere opposition alone to the no platform policy makes one subject to the no platform policy, surely that is an extremist policy in itself. Indeed, I, the author of this piece, would not be given a platform to speak on this basis by the student unions.

The increasing tendency for student unions to censor speakers who may offend them is concerning. Certainly, many may not agree with controversial speakers. However, blocking their free expression not only infringes on free speech but reduces the flow of ideas. It epitomises the undemocratic position that the views of those in powerful positions is more important than the views of others. It is only through allowing controversial speakers, and for people to hear a variety of views that one can be sure they have reached an informed view. Restricting freedom of speech, thus, reveals a lack of respect for the personal autonomy of individuals and their freedom to make up their own minds.

If those who seek to ban controversial speakers instead work to formulate counter-arguments, they will find a richer and more fulfilling public discourse.

'Offensiveness' for the purpose of the so-called 'no-platform' policy is vague and can lead to speakers being arbitrarily cut off from discourse at the behest of those leading student unions. Who are they to say whether something, in particular, is offensive? And even if a view offends a few individuals, it should be through debate that it is countered and not by shutting the debate off.

Question 12

What unstated assumption was made by the author in stating that he would not be given a platform to speak?

A. He would not be allowed to speak by student unions

B. He agrees with Germaine Greer's views

C. He would have the same fate as Peter Tatchell

D. He is a campaigner for gay rights

E. The officer's views on the no-platform policy are reflective of the student unions

Question 13

Regarding freedom of speech?

A. It is a human right

B. Freedom of speech may be countered in circumstances where there are offensive views

C. Peter Tatchell should not have been blocked from speaking

D. Respect for personal autonomy requires freedom of speech

E. There should not have been protests against Germaine Greer's speech

Question 14

Which of the following is an opinion?

A. Tatchell signed an open letter denouncing the no-platform policy

B. 'Offensiveness' has a number of different meanings

C. A number of student unions operate a 'no-platform' policy

D. Students protested against Germaine Greer's talk

E. The increasing tendency for student unions to censor speakers who may offend them is concerning

Question 15

Which of the following follows from the passage?

A. Students are extremists

B. Students are against free speech

C. Many students are against free speech

D. Student unions express the views of the student body

E. Some student unions have been trying to prevent certain individuals from speaking

Passage 4 - Bering Fur Seal Arbitration

The Bering Fur Seal Arbitration was one of the earliest international environmental law cases between the United States and Great Britain. The conflict arose once the US bought Alaska from Russia and thereafter granted a monopoly to the Alaskan Commercial Company through granting the company the exclusive right of sealing on the Pribilof Islands, at the edge of US territorial waters in the Bering Sea. In the 1880s, however, Canadian ships engaged in killing seals in the water, an activity known as pelagic sealing, just outside US waters. This depleted seal stocks inside US waters and on US land, which led to an international conflict between the United States and Great Britain, acting on behalf of Canada.

The United States wanted the upkeep of fur seals and, after the failure of negotiations with Great Britain, an arbitration tribunal was set up to resolve the issue. In order to be successful, the US needed to assert some type of right to the seals outside their territory. Among the arguments advanced by them at the arbitration, it was claimed that the US had the exclusive jurisdiction of the Bering Sea, but this argument was flatly rejected. Alternatively, the US put forward the proposition that they were acting for the benefit of mankind generally to keep up the seal stocks. Additionally, they suggested that sealing on the land, which the US engaged in, was right but that pelagic sealing was illegitimate. This argument, convenient though it was, had never been recognised in any system of law, international or domestic and would have amounted creating new law. Hence, while their counsel displayed ingenuity, they were effectively countered by the counsel for Great Britain, who ended up persuading the tribunal.

In accordance with the terms of the arbitration treaty, the tribunal laid down a number of regulations for the purpose of preserving the fur seal stocks, which were binding on both Great Britain and the United States. The main regulation was the creation of a 60-mile exclusion zone, where pelagic sealing was not allowed except at certain times and in certain ways. One such regulation prohibited the use of firearms and explosives.

Question 16

Why did the US bring the case against Great Britain?

A. Great Britain had responsibility for Canada at the time
B. Canadian ships were killing the US' seals
C. Canadian ships had breached US waters
D. The seal stocks were being depleted
E. As they had a monopoly over their seals

Question 17

Who did the arbitration tribunal find in favour of?

A. The United States
B. Canada
C. Great Britain
D. Alaska
E. No one

Question 18

Why was the US' argument convenient?

A. It fitted in well with their case
B. The US was sealing on land and Canada's was in the water
C. It was not held to be convenient
D. It created a new law
E. It gave them exclusive jurisdiction

Question 19

Which of the following is true?

A. Sealing was prohibited
B. The US could not seal on the Pribilof islands
C. The US were acting for the benefit of mankind
D. Great Britain was more restricted in sealing in the exclusion zone
E. Great Britain lost the case

Question 20

Which of the following would the regulation allow?

A. British ships to seal in the exclusion zone at restricted times
B. US ships to seal in the exclusion zone at restricted times
C. US ships to seal on the land on the Pribilof islands
D. Sealing at all hours, so long as it was without a firearm or explosive
E. No sealing at all

Passage 5 – Footballers' Pay

In 2012-13, the average yearly salary of a Premier League footballer was £1.6m, while at the same time, the average salary in the UK stood at £26,500. The footballers at the top of the tree earn significantly above the average footballer even, with Wayne Rooney on a reported £13 million contract with Manchester United, thereby earning him over 9 times the average every week. This may seem a lot for kicking a ball and running around a field. Footballers tend to enjoy extravagant lifestyles, which have been brought into public focus in recent times given the tough economic conditions that have affected the rest of society.

In 2008, the Labour Government's Sports minister labelled footballers' salaries as obscene, pointing out the difference when compared to the ordinary worker. The Archbishop of York believes that there must be higher taxes for them. What makes a footballer's occupation more worthwhile to society than that of a doctor or nurse, who goes about saving lives? It is indeed arguable that their success is in a big part due to the talent they have from birth, and that is entirely down to luck. It may, thus, seem unfair that footballers earn over a hundred times more than nurses or doctors, many of whom have to work late shifts and make life-and-death decisions and seem to contribute more value to society.

On the contrary, though, the workings of the free market can be used to explain the seemingly excessive salaries of Premier League footballers. Put simply, there aren't many people with the skills of such footballers – they are in short supply – and at the same time, many people – in fact, millions – want to watch them, either at a football match or on television. Consequently, television broadcasters are willing to pay mammoth sums to have the TV rights to football matches and companies are willing to pay millions to sponsor mainstream football teams. These give the teams the enormous purchasing power with which to pay their players large sums. Given that each team wants to win, they will want the best players and thus, will bid against each other for them, thereby inflating their wages even further.

However, if it's just a matter of luck, is it fair that footballers are paid significantly more than doctors? Indeed, supporters of the status quo point to the fact that their respective wages are determined by the market makes it fair – their wage is reflective of their value to society. Millions of people excitedly tune in to watch football and it is, therefore, argued by supporters that society, by this choice, determines the footballer's wage.

On the contrary, though, opponents point out that the market-determined wage (or the market value attributed to one's skills) is mainly determined by luck and not entirely by oneself. Accordingly, the wage differential is not particularly fair given that someone earning a low wage, such as a junior doctor, may well work much harder than a person, such as a footballer, earning a significantly higher wage.

However, it is wrong to focus this debate solely on footballers. It requires significant political discourse as it goes to the heart of a fundamental part of society: the free market system. How, whether and to what extent this should be reformed are matters for careful political debate. The market system does not just lead to footballers enjoying large salaries, but other sports stars, celebrities in general, bankers, lawyers, chief executives etc. earn wages that are significantly above that of the average too. The position of footballers should be seen as not much different from these other professions. It is, thus, incoherent to consider the position of footballers' salaries and not consider society as a whole

Crucially, 45% income tax is already paid on earnings over £150,000 and it is up to democratically elected governments to consider whether taxation suffices or if a wholesale change of the pay system is warranted, such as imposing 'wage-caps'. When considering the latter, however, it is worth bearing in mind that it is not inexorable that the reduction would go to lower earners and not boost the company's own profits.

Question 21

What do the opponents and supporters agree on?

A. What the wage rate for footballers and nurses should be

B. That fairness has different meanings depending on the context

C. That the existing settlement is fair

D. That fairness has a role to play in the wage debate

E. That fairness leads to the current outcome

Question 22

Which of the following is not inconsistent with the arguments of <u>both</u> the opponents and supporters of footballers' wages?

A. Footballers benefit from luck to have their talent

B. Footballers have a fair wage

C. Footballers' wage should be determined by the free market

D. It is unfair that nurses are paid less

E. Footballers should be paid less

Question 23

Which of the following is not inconsistent with the author's argument in the passage?

A. One's hard work is a factor for success

B. It is obvious that the market-based system needs to be overhauled

C. Fairness is a definitive concept

D. Wage differentials reflect hard work

E. Footballers don't pay tax

Question 24

What does the word 'inexorable' mean in this context?

A. Unnecessary

B. Mindless

C. Sporadic

D. Random

E. Unavoidable

END OF SECTION

Section 2

YOU MUST ANSWER <u>ONLY</u> <u>ONE</u> OF THE FOLLOWING QUESTIONS

Question 1

Design an experiment to deduce the sensitivity of a snake's hearing. Explain everything you would do, and your rationale for doing so.

Question 2

"The eternal mystery of this world is its comprehensibility"

To what extent is the world comprehensible?

Question 3

"The greatest obstacle to learning is education"

Argue for or against this statement.

Question 4

Does a vacuum really exist?

END OF PAPER

ANSWERS

Answer Key

Paper A						Paper B					
Section 1A		**Section 1B**		**Section 1C**		**Section 1A**		**Section 1B**		**Section 1C**	
1	B	1	E	1	C	1	B	1	E	1	C
2	D	2	B	2	D	2	D	2	B	2	D
3	E	3	B	3	A	3	E	3	B	3	A
4	C	4	F	4	A	4	C	4	F	4	A
5	C	5	C	5	E	5	C	5	C	5	E
6	C	6	D	6	B	6	C	6	D	6	B
7	E	7	B	7	A	7	E	7	C	7	A
8	C	8	E	8	D	8	C	8	E	8	D
9	D	9	C	9	A	9	D	9	C	9	A
10	D	10	D	10	A	10	D	10	D	10	A
11	C	11	B	11	C	11	C	11	B	11	C
12	C	12	E	12	A	12	C	12	E	12	A
13	D	13	B	13	C	13	D	13	B	13	C
14	B	14	C	14	D	14	B	14	C	14	D
15	B	15	D	15	A	15	B	15	D	15	A
16	E	16	F	16	A	16	E	16	F	16	A
17	C	17	E	17	A	17	C	17	E	17	A
18	C	18	A	18	B	18	C	18	A	18	B
19	B	19	F	19	C	19	B	19	F	19	C
20	C	20	C	20	B	20	C	20	C	20	B
21	E	21	G	21	C	21	E	21	G	21	C
22	C	22	C	22	D	22	C	22	C	22	D
		23	B	23	B			23	F	23	B
		24	B	24	D			24	B	24	D
		25	C					25	C		
		26	C					26	C		
		27	A					27	A		
		28	B					28	B		
		29	C					29	C		
		30	E					30	E		

Mock Paper A Answers

Section 1A

Question 1: B
James runs 26.2 seconds, which is outside the qualifying time, therefore he does not qualify

Question 2: D
5.6/7 gives the unit price of 80p – this equals a packet of crisps. Multiplying this by 2 gives the sandwich and by 4 gives the watermelon price of £3.20

Question 3: E
Jane leaves at 2:35pm and arrives at 3:25pm, taking 50 minutes. Sam's journey takes twice as long, so leaving at 3:00pm it takes 100 minutes, giving an arrival time of 4:40pm

Question 4: C
After the donation, Sam has eight sweets. Therefore, Hannah had 16 sweets after the transaction and hence 13 sweets before

Question 5: C
Find original pay: £250/0.86 = 290 basic original pay. Add the rise: (290 x 1.05) + 6 = £311 new basic pay. Subtract the income tax at 12% = 311 x 0.88 = £273 new pay rate

Question 6: C
Given the first cube is a white cube, you are drawing from one of three boxes, boxes A, C or D. Boxes C and D will have just had their only white cube removed, whereas box A will have one white cube remaining. Therefore, the probability of drawing a second white cube is $^1/_3$, thus the probability of non-white (i.e. black) is $^2/_3$.

Question 7: E
This is a simultaneous equations question. 500 + 10(x – 80) = 600 + 5x; true when x ≥ 80.
500 + 10x – 800 = 600 + 5x
» 5x = 900
» x = 180

Question 8: C
If eating more slowly caused a reduction in the time available to work, the candidate might be less productive.

Question 9: D
This is a LCM question. We need to find the lowest common multiple of the song lengths. The LCM of 100, 180 and 240 is 3,600 seconds – equal to 60 minutes. For ease of arithmetic, you may choose to work reduce all numbers by a factor of 10.

Question 10: D
The journey is 3 hours and 45 mins, minus a 14-minute break gives 3hrs 31 mins travel time, or 211 minutes. Therefore, the average speed is 51mph, or 82kmh by using the stated conversion factor.

Question 11: C
The mean guess is £13.80, which is £5.80 too high

Question 12: C
The overall error for respondent 3 is £13, which is the least

Question 13: D

Scale back and forth from known quantities. Country B has 32.1m so Country D has 38.6m people.

Question 14: B

Country B has 32.1m people. Therefore $45 x 32.1m = $1.45bn

Question 15: B

The average speed is 24mph, independent of distance travelled as it cancels. Imagine this covers a set distance of say 30 miles. It will take 1 hour on the way and 1.5 hours on the way back. 60/2.5 = 24. This is true of all distances; the ratio is the same.

Question 16: E

None of the above can be reliably deduced from the passage alone

Question 17: C

Imagine the toothpaste costs 100p originally, and follow the price through. It rises by 80% to 180p, then is reduced by 50% to 90p. Three tubes are purchased for the price of 2 (i.e. 180p), therefore the cost per unit is 180/3 = 60p. 60p = 60% x 100, the original price

Question 18: C

Statement C is the only one referring to the potential outcome of solving crimes faster, thereby providing a plausible mechanism for a reduction in cybercrime rates

Question 19: B

The passage suggests that the attacks were carried out by extra-terrestrial beings. Though the supposed UFO sightings have rational explanations, the writer feels this is insufficient to dismiss his idea.

Question 20: C

The initial argument suggests that two things must be present for an action to happen. If only one is absent, the action cannot happen. Argument C has the same form, the others do not.

Question 21: E

Growing vegetables needs several positive traits. The passage does not tell us which is the most important or most commonly lacked skill, only that more than one skill is required for success.

Question 22: C

Joseph does not have blue cubic blocks, since all his blue block are cylindrical.

END OF SECTION

Section 1B

Question 1: E
Haemoglobin is contained within red blood cells and is not free in the blood. Additionally, as a protein it is too large to normally pass through the glomerular filtration barrier. All the other substances are freely filtered.

Question 2: B
More sodium than potassium must move inwards in order to depolarise the membrane. This results in a net movement of positive charge into the cell, causing a depolarisation.

Question 3: B
Equate the volume with the surface area in the proportion instructed by the question. $3(^4/_3\pi r^3) = 4\pi r^2$, simplifies to $r = 1$.

Question 4: F
A polymer consists of repeating monomeric subunits. Polythene consists of multiple ethenes; glycogen of glucose; collagen of amino acids, starch of glucose; DNA of nucleotide bases, but triglycerides are not composed of monomeric subunits.

Question 5: C
Potassium is the most reactive as it has a single donor electron and is bigger than sodium (next most reactive). Copper is the least reactive of the metals listed.

Question 6: D
Diastole is the relaxation phase of the cardiac cycle. In diastole the pressure in the aorta decreases as the contractile force from the ventricles is reduced and the blood expelled dissipates into the arteries. Therefore, E is the only untrue statement about diastole.

Question 7: B
Competitive inhibition occurs when the inhibitor prevents a reaction by binding to the enzyme active site. Hence, a higher concentration of the substrate can result in the same overall rate of reaction. i.e. the substrate outcompetes the competitor.

Non competitive inhibition is where the inhibitor binds to the enzyme (not at the active site) and prevents the reaction from taking place. Increasing the substrate concentration therefore does not increase the reaction rate i.e. the substrate cannot outcompete the competitor as the enzymes are disabled and the competitor is not binding to the active site.

In this graph, line 1 shows the normal reaction without inhibition, line 2 shows competitive inhibitor and line 3 shows non-competitive inhibition.

Question 8: E
Nucleic acids are only found in the nucleus (DNA & RNA) and cytoplasm (RNA). They are not a component of the plasma membrane, whereas the other molecules are.

Question 9: C
Multiply together the number of international airports by the annual flights per airport. The annual flights per airport is taken by multiplying the number of flights per hour by the number of hours in a year (365 x 24).

Question 10: D
Add the first and last equations together to give: $2F = 4$, thus $F = 2$.
Then add the second and third equations to give $2F - 2H = 5$. Thus, $H = -0.5$
Finally, substitute back in to the first equation to give $2 + G - 0.5 = 1$. Thus, $G = -0.5$
Therefore, FGH = 2 x -0.5 x -0.5 = 0.5.

Question 11: B
Note that the units are the same (M = moldm^{-3}), only the orders of magnitude are different. Convert the orders of magnitude to discover a 10^6 difference with more chloride than thyroxine

Question 12: E
The way to solve this is to break the calculation down into parts, almost working backwards. The number of seconds is given by: 60 x 60 x 24 x 7 x 66:
= (10 x 6) x (12 x 5) x (4 x 6) x 7 x (11 x 6)
= 1 x 4 x 5 x 6 x 6 x 6 x 7 x 10 x 11 x 12
= 1 x 4 x 5 x (3 x 2) x 6 x 7 x 10 x 11 x 72
= 1 x 2 x 3 x 4 x 5 x 6 x 7 x 8 x 9 x 10 x11

Question 13: B
Glycogen is not a hormone, it is a polysaccharide storage product primarily found in muscle and the liver.

Question 14: C
Remember the interior angles of a pentagon add up to 540° (three internal triangles), so each interior angle is 540/5 = 108°. Therefore, angle **a** is 108°. Recalling that angles within a quadrilateral sum to 360°, we can calculate **b**. The larger angle in the central quadrilateral is 360° – 2 x 108° (angles at a point) = 144°. Therefore, the remaining angle, **b** = (360 – 2(144)]/2 = 36°. The product of 36 and 108 is 3,888°.

Question 15: D
Working by orders of magnitude, multiply all the bacteria tested on the numerator and the number of resistances on the denominator. This gives an order of 10^{25}, which is the solution.

Question 16: F
None of the above, they are all true facts about digestion.

Question 17: E
Blood flow to the kidneys is not exercise dependant. It is constant. Overall cardiac output increases, there is more blood flow to the muscles to fuel them and to the skin to help lose excess heat. Blood flow to the gut decreases to increase availability to muscles. Blood flow to vital organs such as the kidney and brain is constant.

Question 18: A
Since A-T and C-G are the DNA base pairings, 29.6% Adenine implies 29.6% Thymine as well. Therefore, the remaining 100 – 59.2 = 40.8% is shared between Guanine and Cytosine equally, so there is 20.4% cytosine.

Question 19: F
Structure A is the right semi-lunar valve, the pulmonary valve. It opens in systole to allow flow of blood from the right ventricle into the pulmonary artery and to the lungs. It closes in diastole to ensure the right ventricle fills only from the right atrium, maintaining a one-way flow of blood. Therefore, F is true, it opens when the right atrium is emptying. None of the other statements are true.

Question 20: C
Since CO binds to the oxygen binding site of haemoglobin, it reduces oxygen binding. Therefore, the blood is less oxygenated, so heart rate must increase, as more blood needs to flow to deliver the same amount of oxygen.

Question 21: G
The replacement of dying, damaged, and lost cells, the growth of the embryonic cell to a multicellular organism, and asexual reproduction are the three main reasons why cells divide through mitosis

Question 22: C

The formula for the sum of internal angles in a regular polygon is given by: (180)n-2. Thus:

180n − 360 = 150n

3n = 36

n= 12

Question 23: B

Start by multiplying each term by ax to give: a(y+x)=x^2+a^2

Expand the brackets: ay+ax=x^2+a^2

Subtract ax from both sides: ay=x^2+a^2-ax

Lastly, divide the both sides by a to get: $y = \frac{x^2+a^2-ax}{a}$

Question 24: B

The shortest distance between points A and B is a direct line. Using Pythagoras:

The diagonal of a sports field $= \sqrt{40^2 + 30^2} = \sqrt{1,600 + 900} = \sqrt{2,500} = 50$.

The diagonal between the sports fields $= \sqrt{4^2 + 3^2} = \sqrt{16 + 9} = \sqrt{25} = 5$.

Thus, the shortest distance between A and B $= 50 + 5 + 50 = 105\ m$.

Question 25: C

Let $y = 1.25 \times 10^8$; this is not necessary, but helpful, as the question can then be expressed as: $\frac{100y + 10y}{2y} = \frac{110y}{2y} = 55$

Question 26: C Chemical reactions take place in the cytoplasm, and the mitochondrion is the site for aerobic respiration releasing energy. The lack of a cell wall means that this is an animal cell.

Question 27: A

Solve simultaneously:

$y = 2x - 1 = x^2 - 1$

$If\ y = 0$:

$2x - 1 - x^2 + 1 = 0$

$2x - x^2 = x(2 - x) = 0$

Thus, x = 2 and x =0

There is no need to substitute back to get the y values as only option A satisfies the x values.

Question 28: B

The ruler and the cruise ship look to be the same size because their edges are in line with Tim's line of sight. His eyes form the apex of two similar triangles. All the sides of two similar triangles are in the same ratio since the angles are the same, therefore:

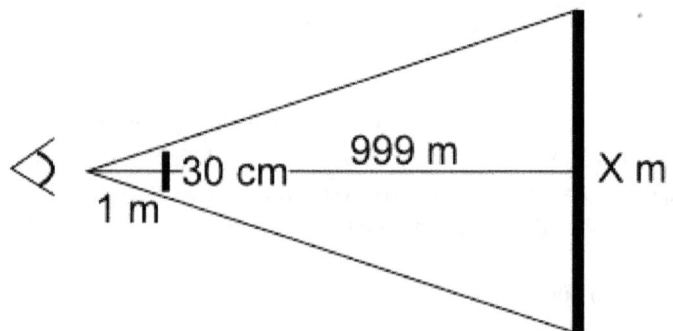

$\frac{30\ cm}{X\ m} = \frac{1\ m}{1\ m+999\ m}$

Thus, $X\ m = 1000\ m \times \frac{30\ cm}{1\ m}$

$1000 \times 0.3 = 300\ m$

Question 29: C

Bob = B, Kerry = K, Son = S.

B = 2K
K = 3S
B + K + S = 50

$50 = 2K + K + K/3 = 6K/3 + 3K/3 + K/3 = 10K/3$
K = 15
B = 30
S = 5

So: B – S = 30 – 5 = 25.

Question 30: E

Blood pressure in the aorta is the highest of any vessel in the body, as blood has just been ejected from the left ventricle to go to the body.

END OF SECTION

Section 1C

Passage 1

Question 1: C

Women in Illinois, not across the USA, were subject to the law, and the passage does not state either a change in fashion or actual arrests, only the potential for arrests.

Question 2: D

The pulling out of feathers from live birds was seen as the negative alternative to using osprey feathers.

Question 3: A

They could be possessed only 'in their proper season'.

Question 4: A

The problem cited is that the article was already in use in the clothing of numerous military men. The authority of the princess/sexist politics does not feature in the passage, and 'D' is patently false.

Question 5: E

None of those are precluded, as only 'harmless' and 'dead' birds (in their entirety) were prohibited. Wearing a living bird was not explicitly banned.

Passage 2

Question 6: B

He thought it was 'a pity that only rich people could own books', and from this, he 'finally determined to contrive' of a new way of printing. The passage does not state that he wished to make money, found books too expensive to get a hold of, or was impatient himself when it came to the production of books.

Question 7: A

The need to be careful is mentioned, as is the fact that the process takes a long time both in creating the block and due to the fact one block can only print one page. That it may tire a carver to make the block is possible, but it is not cited in the passage.

Question 8: D

The statement says it is 'very likely' he was taught to read but is not definite. The fact that his father comes from a 'good family' does not mean he is a member of the aristocracy, necessarily. Though block printing was used as the boy grew up, it does not state this was the most popular process. The mention of Gutenberg's family's 'wealthy friends' indicates that they were sociable.

Question 9: A

The paper was laid on top of the block, not underneath.

Question 10: A

There is nothing written in the passage praising the craftsmanship of manuscripts. The appropriateness of the titles for both book production processes is explained, and the 'wealthy friends' are described as a source to borrow books from, and, thus, a way of expanding one's reading. The boy was 'very likely' taught to read as well which implies that he was probably educated.

Passage 3
Question 11: C

Nowhere does it state that all European countries have similar creatures (though certain types can be found in both British Isles and Norway) nor does it state that the array of animals is limited to this one nation. Sharing animals and birds does not necessitate sharing geographical features, but it is said that a country with forest and moorlands is likely to have a variety of birds and animals, so one can see the link between forests and creatures.

Question 12: A

There was a time when the English dreaded wolves and bears, but that indicates the past, or at least does not include the present. Norwegians being superior is not suggested here.

Question 13: C

Bears are called destroyers, which is sufficient to conclude that they cause damage.

Question 14: D

They are ruthlessly hunted by farmers in country districts, but numerous only in the forest tracts in the Far North.

Question 15: A

The word 'fortunately' implies that it is good that the wolves are no longer central. The children are under no threat, as the threat of wolves belongs to a bygone time, there is no mention of regret that such a time is gone and Norsemen are not presented in the above passage as having respect for Nature, but instead, they are said to interfere with it through hunting and driving wolves farther afield from their current homes.

Question 16: A

This is not based on evidence and can vary from person to person. It can't be tested as being true or false and, therefore, is an opinion.

Passage 4
Question 17: A

Most requires over half by definition, and "most" of the people living in this area were the descendants of immigrants who moved to the country a "full century ago".

Question 18: B

Hall only makes a claim for New England, not the entirety of America, being the descendants of 20,000 immigrants. The 'one million' figure comes from Franklin, not Hall. Less than 80,000 ("under" 80,000) people led to the population boom of one million. One million is over ten times more than 80,000, so "B" is correct.

Question 19: C

It is said to be "distinct" to older aristocracy "of the royal governor's courts". It is not similar to any European aristocracy. There is no specific reference to it not being a system based on lineage.

Question 20: B

"A", "C" and "D" are cited in the passage (the journey took 'the better part of the year', it was 'hazardous' and 'expensive'), whereas 'B' is not referenced at all.

The word 'reference' best fits in as the author is referring to the lack of citation of Shakespeare and the Puritan poet Milton.

Passage 5

Question 21: C

It would be a massive assumption to state that just because two characters in a book are 'vicious', all of them will be, so 'A' is not necessarily correct. 'B' also believes in a despair that is described to belong to the Comedian, but not Rorschach. The argument of the passage is that 'D', which Moore may believe, is not the case - the beloved character is not simply worshipped for his violence, but for his belief in justice. 'C' is correct, as Moore describes how he did not wish for Rorschach to be a favourite character, but rather a warning.

Question 22: D

Rorschach is considered as a hero, despite using violence. Therefore, A and B are wrong. The author then goes on to highlight that the issue with the Comedian is that he has no purpose but at least Rorschach does have a purpose, even if he does use violence. Accordingly, D naturally follows from this. C and E do not follow from what is discussed.

Question 23: B

He does not mention madness ('D'), or invoke shame ('C'), or simply state it is good to be good ('A') - specifically, he states we must act as if the world is 'just', even when it is not, in order to attain dignity.

Question 24: D

No value judgment is made regarding violent actions or on the Comedian's jokes so 'A' and 'B' are false. The passage also acknowledges Rorschach's violence, showing 'C' is wrong but does state that his actions are due to the fact he believes that he is acting in the name of justice, which lends him an ethical justification for his actions.

END OF SECTION

Section 2

1. *"Strive not to be a success, but to be of value"*

To what extent is it possible to be "a success", but to have little value?

Introduction:
- Definitions of success and value should be given, with a brief exploration of what is implied by the two terms and any connotations.
- Success could be considered to be defined through achievements or status, whilst value is defined on a more emotional or social level with the influence that people have on the situation around them.
- The introduction should begin to explore where there are overlaps between the two terms and where they could be seen as standing alone. This could be done by general statements or through specific examples, so long as valid ideas are introduced on both sides.

Possible arguments for success without value:
- In the context of money – a person could have a successful career defined by the amount of money that they make, but may not have been a positive influence on those around them. You could consider the extent to which the definitions may get confused with worth but doesn't mean it (e.g. she's worth £xxx as a term for success)
- In the context of academic achievement – just because you have been successful with grades at school, for example, doesn't mean you were necessarily valued in the setting or by those around you.
- In the context of gambling or competitive sports/games – just because a person has been successful in 'winning' doesn't necessarily mean that they were valued as a player or by others around them.

Possible arguments for success and value being integrated:
- Is success a value in itself? Could this be in the eye of the beholder?
- In terms of gaining power or authority – you could choose to explore whether figures in power such as a manager, politician etc. must be valued by somebody to reach this status?
- In particular careers such as teaching or charity work – in order to be successful must you be doing something that is valued by somebody?
- On an emotional level, could you be considered a 'successful' friend or family member if you are valued within this setting? This may be particularly relevant using the example of a mother figure.
- You may wish to put forward the idea that you can be valuable without success, but does this work the other way around?

Conclusion:
- Include a summary of all points given, and refer back to your original definition.
- The conclusion should reach a clear and logical solution – you could choose to completely decide that it is impossible, but most probably will reach a compromise which encompasses the idea that value is distinct from success, but the terms can be used interchangeably in particular situations.

2. Is the media a positive or negative influence on scientific understanding?

Introduction:
- Introduce or illustrate the ways in which the media are involved in our scientific understanding – you could describe what ways we might have access to the information, such as newspapers, TV.
- You may give reference to the particular areas of scientific interest in the media, such as health and global warming.
- Introduce both sides of the argument – the possible positive influences and possible negative ones (ideas below).

Possible positive influences:
- Action- It could be argued that we wouldn't be aware of some of the biggest scientific issues needing tackling if it wasn't for the involvement of the media, and therefore the effect must be positive if we are becoming aware of pressing problems (e.g. obesity)
- Access - The media is a way through which the world of scientific research integrates with common people, who would not otherwise recognise the importance of science.
- Description – the media may be able to explain ideas in a more easily understandable manner, such as by identifying the key trends and emphasising the important facts, which makes science more understandable to those with less knowledge.
- It could be argued that any awareness of science is positive for those who would not have any interest otherwise.
- Examples of specific situations – e.g. in the banning of CFC sprays.

Possible negative influences:
- Simple facts may get blown out of proportion or taken out of context to such an extent that they could no longer even be considered factual or valid.
- False or flawed statistics are often given, taken from unreliable sources, which lead to confusion and the wrong message being portrayed.
- Oversimplification of ideas – everything is dumbed down and will not give the full story.
- Can create a biased public opinion – a conclusion may be reached which does not consider all particular explanations for a phenomenon.

Conclusion:
- Include a summary of all points made and give a balanced overview of both sides of the argument.
- You may wish to reach a unanimous decision on the effect, making it clear that this is a personal opinion. Or you may wish to come to a compromise where the media are a positive in some situations or to some extent, but after a particular point or when taken too far, this effect can become negative.

3. *"Why tell the truth if a lie is better for all concerned?"*

In what circumstances can dishonesty be justified?

Introduction:

- It may be important to define the term 'lie' as a statement or implication that is directly opposed to the truth. It is also important to note that, in this context, a lie is a deliberate dishonest remark (not an accidental incorrect statement)
- You may wish to distinguish between lying as deliberately telling something that is factually incorrect, and deliberately avoiding or obscuring the truth.
- Introduce the idea that dishonesty is considered immoral and 'wrong' – the basic social rules are not to lie, children are taught from a young age that lying is bad behaviour.
- Begin to introduce your arguments in favour of lying in particular situations – the points you wish to explore later.

You may wish to structure this essay slightly differently to this mark scheme; through exploring a circumstance where it may be considered justifiable to lie, and then look at the alternative perspective arguing that it is not justified. Alternatively, it would be just as effective to make points for and against, as below.

Possible arguments for:

- In the case of 'white lies' – to avoid discouraging or offending people e.g. "Your hair looks nice"
- In self-defence or when lying may save a number of lives – e.g. when people are taken hostage for their beliefs/ethnicity/race
- When talking to children about situations they wouldn't otherwise understand, especially when they ask difficult questions.
- To avoid an unnecessary, difficult conversation with someone who doesn't need to know something detailed – e.g. if they ask, "How are you?" not giving them a truthful answer may avoid a difficult conversation.

Possible arguments against:

- Creates a lack of trust and a lack of stability, feelings of guilt on both sides.
- Religious/foundations – 'do not lie' is one of the Ten Commandments.
- You could argue that lying will always cause more problems later down the line, even if making things easier in the short term.
- Advocating lying in some situations makes it easier to lie in other situations and creates a perpetual problem, it's impossible to draw a line other than to just say one should not lie as a rule.

Conclusion:

- Include a summary of all points made, and give a balanced overview of both sides of the argument.
- You need to reach a decision which illustrates if, and when, dishonesty can be justified. It's OK to also reach the conclusion that it never can be, so long as this is logically stated.

4. "Science is a nothing more than a thought process"

What actually is science and how is it of value to us?

This essay is very open-ended: it is possible to take this question in any direction you feel appropriate depending on personal knowledge and interest so this mark scheme is by no means the only way of approaching this question – so long as a convincing and balanced argument is given and both elements of the question are considered, the answer will be credible.

Introduction:
- Include a definition of science – a systematically organized study or body of knowledge of the physical and natural world.
- Begin to introduce some of the key features of science.
- Begin to introduce some of the values that science has for society.

Possible points to consider – you want to contain a balance of defining and exploring the key features of science, and exploring the value of science within our society:
- Science is experimental – involving accurate and detailed study.
- Science could be argued to be physical and measurable- using known concepts in the real world that can be objectively identified.
- Science is adaptable and ever changing – according to research and current thought.
- Science could be considered to be held within the thoughts and brains of society – if there weren't people exploring it or interested in it, would it really exist?
- Science involves logical and critical thought – in that sense it could be considered to be merely a thought process – but you may wish to explore the idea that it couldn't exist without physical things to measure in themselves.
- Valuable for the evolution of society – as we learn more about the world and about ourselves, this helps change to occur which is vital.
- Provides concrete fact and stability.
- Gives a purpose to life, which brings fulfilment and contentedness.
- Brings explanation for some things that can't be explained through simple observations, which allows a firmer foundation and an answer to bigger, crucial questions.
- Helps us to gain more of an understanding of the world we are in – bringing a stronger grasp on reality and greater knowledge and insight.
- Medical reasons – drugs, medicines, health.
- Safety – e.g. monitoring volcanoes, predicting earthquakes etc.

Conclusion:
- Summarize the key ideas explored previously – the ideas associated with science and the values it may hold.
- Ensure to link back to the original statement – is it within our minds and merely a thought process – and does that diminish its' value?

END OF PAPER

Mock Paper B Answers

Section 1A

Question 1: C
This question has to be worked through in stages. To begin with, adjust his weekly pay to an annual salary, (560 x 52 = £29,120). Accounting for tax, his annual salary is 29,120/0.9 = 32,355. To account for the pay cut, 32,355 x 0.95 = £30,737. To deduct tax, subtract 20% of 20,737 (his taxable income) to give 4,147 tax. Subtracting the new tax from his new overall salary gives 26,589. Add on the new bonus of 90 per month to give £27,669, or £27,700 when rounded.

Question 2: B
The total saving on the final booking relative to the first is £230, but the cost of two cancellations must be deducted (£90) giving a total saving of £140.

Question 3: B
There are originally no odd numbered balls in Bag A. But as a result of the transfer, there could be an odd ball in Bag A. Therefore, the probability of drawing an odd ball is found by multiplying the probability of selecting the new ball ($^1/_5$) by the probability that that ball is odd ($^2/_5$ – given by adding the one odd ball in the bag originally to the odd ball introduced) giving a probability of $^2/_{25}$ that the selected ball from Bag A is odd.

Question 4: B
Assume the price of bread is 100p. 100 x 1.4 x 0.8 = 112p after the subsidy. The cost of three loaves is therefore 336p (divided four ways this equals 84p per loaf)

Question 5: D
At 2120hrs, the minute hand is pointing to 4 and the hour hand is pointing one third of the way past 9 towards 10. 360°/12 = 30° – this is the number of degrees per hour division. Between the two hands then, there are 5-hour divisions plus an extra $^1/_3$. Therefore, the angle is (30x5)+(30/3) = 160°

Question 6: B
There is a 3l and 5l bucket – therefore 4 litres can be measured from the difference between the buckets as follows. Fill the 5l bucket, decant 3l into the smaller bucket and then you are left with 2l in the large bucket. Pour this into the tank. Repeat the process again, decanting the remaining 2l into the tank once again to make 4l in total. The first time, 5 litres was required. The second time, the 3 litres from the second bucket could be tipped back into the 5l bucket, and then filled up with fresh water to measure the final 2 litres in. Therefore 4 + 3 = 7 litres of water are sufficient to fill the tank with 4l.

Question 7: D
To answer this question, make a timeline showing the locations of the different genres of books. Place each book on the timeline as appropriate, making sure to indicate where more than one location is a possibility. From that, you will see that literature books are located to the right of engineering. This is true since they are to the right of art (which we know is right of mathematics (and therefore engineering, since the run between the sciences is uninterrupted)). The other statements, whilst potentially true, cannot be deduced for certain.

Question 8: C
The passage tells us that brand new cars lose value quickly, despite the car being virtually unchanged. Therefore, in the absence of any contradictory information, it is reasonable to conclude that buying second hand cars is a wise choice.

Question 9: B
First, calculate how many bottles are sold. 2000 – (2000x0.9x0.8) = 560 bottles. Then divide the total profit by the number of units to give the profit per unit, which comes to 11200/560 = £20 per bottle.

Question 10: C

If ocean exploration has led to the discovery of many useful drugs, it could easily be said that it has benefitted many people in trouble. Whilst it might be cheaper than space exploration, the two are entirely different, and therefore people's views on the cost-effectiveness of space exploration cannot be directly compared to their views on ocean exploration. The other responses do not address the effect on people in trouble.

Question 11: E

The definition of timelessness requires something to be tested by time. Something that modern furniture cannot fulfil. Therefore, statement E expresses a significant flaw in the reasoning. The other statements do not refer to the 'timelessness' aspect of furniture, therefore they are not directly relevant to the argument.

Question 12: A

The passage talks about the benefits of drinking red wine, not about living near to vineyards. The passage does not state that Italians drink more wine than Germans, therefore the assumption that they do is central to the argument.

Question 12: C

Tom arrives at 1620, and leaves 45 mins after Jane leaves. Therefore, he also leaves 45 mins after Hannah leaves, since Jane and Hannah leave together. Since his journey is 10 mins faster than Hannah's, he arrives only 35 minutes after Hannah arrives (which happens to be 1620). Therefore, Hannah arrives 35 minutes earlier than this, at 1545. Since she left at 1430, her journey took 75 minutes. Jane's journey took 40% longer (1.4 x 75 = 105 minutes). Therefore, leaving at the same time as Hannah, 1430, Jane arrived 105 minutes later at 1615.

Question 12: B

This is a simultaneous equations question. Let x be the number of standard tickets sold, and y be the number of premium tickets sold.
Therefore: $x + y = 600$; $10x + 16y = 6,600$
$x = 600 - y$ » substitute: $10(600 - y) + 16y = 6600$
$6y = 600$; $y = 100$, therefore 100 premium tickets were sold.

Question 13: C

Between 20th January and 23rd May, there are 94 days. In 94 days, the moon makes $94/28 = 3.36$ orbits. This is equal to $3.36 \times 360° = 1210°$

Question 14: E

In question 14, you are looking for a strong opposition to the proposition that students at drama academies are not taught well academically. The strongest opposition would be evidence that such students perform academically well in some objective measure. Evidence of significantly above average GCSE results provides this.

Question 15: D

You should definitely draw this one out on paper. Trace out the paths and you find that both people have a net displacement of 11km to the North. Therefore, since Anil is only net 2km East, and Suresh is 17km East of the starting point, there is a 15km separation between them

Question 16: E

If three times the final amount of concrete is ground off by the builder, three quarters of the original thickness is removed, hence one quarter remains. $14/4 = 3.5$cm

Question 17: A

Walking at 4mph, 3 miles takes ¾ hour = 45 mins. Adding the 5-minute stop, Chris will arrive at 1820, since he set off at 1730. At 24mph, 6 miles takes ¼ hour, 15 mins. Therefore, setting off at 1810, Sarah will arrive at Laura's at 1825. Therefore, Chris arrives 5 minutes earlier than Sarah.

Question 18: D

The passage tells us that illegal downloads are causing harm to the music industry. Whilst it gives an example, this does not mean the stated example is the principal issue. The conclusion that best fits the passage as a whole is to say illegal downloading is more harmful than many people think, given their willingness to undertake it.

Question 19: E

First, calculate the amount of water needed for each type of fire. Typical house fires require 20,000 litres, whereas garden fires usually need only 10,000 litres. Therefore, all statements are correct EXCEPT E. Three houses and ten garden fires require 160,000 litres to extinguish, not 140,000.

Question 20: E

Calculate the distance travelled during each component of the journey, then add them together.
$(20 \times 30 = 600m, (30 + 20)/2 \times 5 = 125m, 30 \times 20 = 600m, 30/2 \times 10 = 150m)$.
Adding the distances together gives 1475m.

END OF SECTION

Section 1B

Question 1: B
Natural selection favours those who are best suited for survival – this can mean faster and stronger organisms, but not always. For example, snails are pervasive, despite being weak and slow. Variation can arise due to both genetic and environmental components.

Question 2: D
The enzyme amylase catalyses the breakdown of starch into sugars in the mouth (1) and the small intestine (5).

Question 3: E
Whilst there is some enzymatic digestion in 1 and 3, the vast majority occurs in the small intestine (5). The liver facilitates digestion via the production of bile, and the large intestine is primarily responsible for the absorption of water.

Question 4: D
The energy in a nuclear bomb comes from $E = mc^2$. When two nuclei fuse, the combined mass is slightly smaller than the two individual nuclei, and the mass lost is converted to energy according to Einstein's equation. Fusion releases much more energy than fission, as in the sun, and humans cannot harness this energy yet. Uncontrolled fission causes the explosion in an atom bomb and is created by a neutron-induced chain reaction. In power plants these neutrons are tightly controlled, so as not to overload the reactors and cause an explosion.

Question 5: B

$$\left(\frac{T}{4\pi}\right)^2 = \frac{l(M + 3m)}{3g(M + 2m)}$$

$$\frac{T^2}{16\pi^2} \times \frac{3g}{l} = \frac{M + 3m}{M + 2m}$$

$$3gT^2(M + 2m) = 16l\pi^2(M + 3m)$$

$$3gT^2M + 6gT^2m = 16l\pi^2M + 48l\pi^2m$$

$$6gT^2m - 48l\pi^2m = 16l\pi^2M - 3gT^2M$$

$$m(6gT^2 - 48l\pi^2) = 16l\pi^2M - 3gT^2M$$

$$m = \frac{16l\pi^2M - 3gT^2M}{6gT^2 - 48l\pi^2}$$

Question 6: A

Replotting the genetic diagram with genotype information produces the diagram:

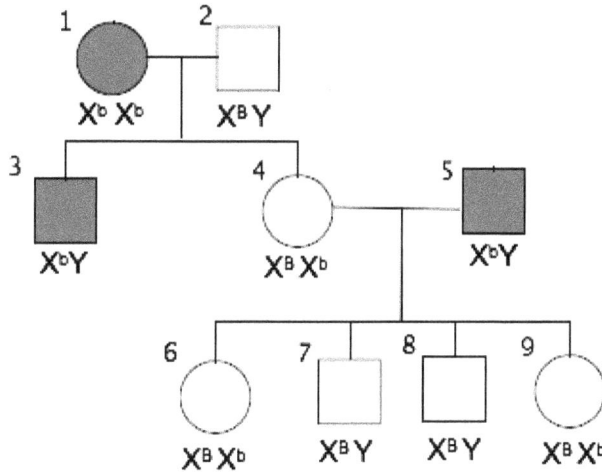

If squares were female, all of 5's circular male offspring would be affected. Circles must be females, so 1 must be homozygous recessive.

Question 7: D

The genotype of individual 6 must be $X^B X^b$, and 8 $X^B Y$. Plotting the information in a Punnett square:

		Individual 6 (Female carrier)	
		X^B	X^b
Individual 8 (Unaffected male)	X^B	$X^B X^B$	$X^B X^b$
	Y	$X^B Y$	$X^b Y$

The progeny produced are 25% $X^B X^B$ (homozygous normal female), 25% $X^B X^b$ (heterozygous carrier female), 25% $X^B Y$ (normal male) and 25% $X^b Y$ (affected male). So, the chance of producing a colour-blind boy is 25%.

Question 8: B

The mean is the sum of all the numbers in the set divided by the number of members in the set. The sum of all the numbers in the original set must be: 11 numbers x mean of 6 = 66. The sum of all the numbers once two are removed must then be: 9 numbers x mean of 5 = 45. Thus, any two numbers which sum to 66 – 45 = 21 could have been removed from the set.

Question 9: E

From the rules of angles made by intersections with parallel lines, all of the angles marked with the same letter are equal. There is no way to find if d = 90°, only that b + d = c = 180° − a = 135°, so b is unknown.

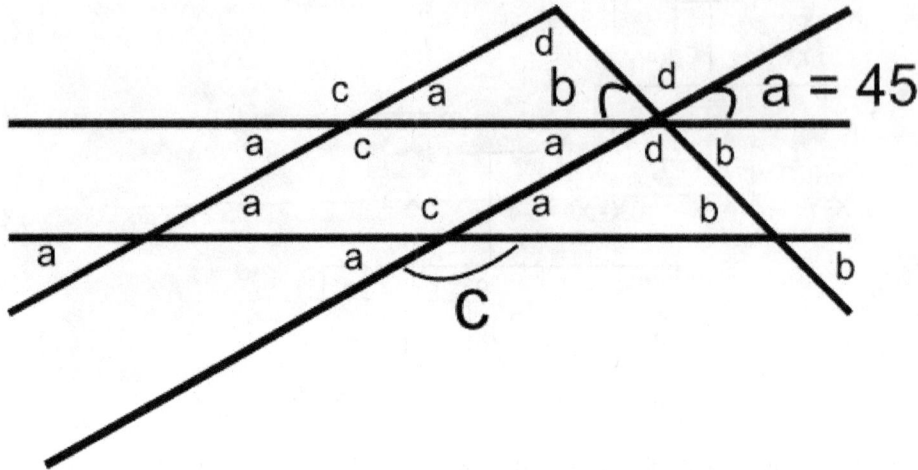

Question 10: B

Let $y = 3.4 \times 10^{10}$; this is not necessary, but helpful, as the question can then be expressed as:

$$\frac{10y + y}{200y} = \frac{11y}{200y} = \frac{11}{200} = \frac{5.5}{100}$$

$$= 5.5 \times 10^{-2}$$

Question 11: B

As the known parent has both recessive genotypes, it can only have the gametes, y and t. The next generation has a phenotypic ratio of 1:1:1:1. As both recessive and dominant traits are present in the progeny, the unknown parent's genotype must contain both the recessive and dominant alleles. Hence the unknown parent's genotype must be YyTt as this would produce the gamete combinations of YT, Yt, yT and yt, which when combined with the known yt gametes would result in YyTt, Yytt, yyTt and yytt in equal ratios.

Question 12: D

The possible genotypes are: YYTT (yellow, tall), YyTT (yellow, tall), yyTT (green, tall), YYTt (yellow, tall), YYtt (yellow, short), YyTt (yellow, tall) Yytt (yellow, short), yyTt (green, tall), yytt (green, short).

Thus, 9 different genotypes and 4 different phenotypes are possible.

Question 13: C

Transform all numbers into fractions then follow the order of operations to simplify. Move the surds next to each other and evaluate systematically:

$$= \left(\frac{6}{8} x \frac{7}{3} \div \frac{7}{5} x \frac{2}{6}\right) x \frac{4}{10} x \frac{15}{100} x \frac{5}{100} x \frac{5}{25} x \pi x \left(\sqrt{e^2}\right) x e\pi^{-1}$$

$$= \left(\frac{42}{24} \div \frac{14}{30}\right) x \frac{4 \, x \, 3 \, x \, 25}{10 \, x \, 20 \, x \, 100 \, x \, 25} x \pi x \pi^{-1} x e^{-1} x e$$

$$= \left(\frac{21}{12} \div \frac{7}{15}\right) x \frac{12}{200 \, x \, 100} x \frac{\pi}{\pi} x \frac{e}{e}$$

$$= \left(\frac{21}{12} x \frac{15}{7}\right) x \frac{3}{50 \, x \, 100}$$

$$= \frac{45}{12} x \frac{3}{5000}$$

$$= \frac{9}{4} x \frac{1}{1000}$$

$$= \frac{9}{4000}$$

Question 14: D

Whilst getting vitamins, killing bacteria, protein synthesis, and maintaining cellular pH and temperature are all important processes that require a blood supply, the MOST important reason for having a blood supply is the delivery of oxygen and removal of CO_2. This allows aerobic respiration to take place, which produces energy for all of the cell's metabolic processes.

Question 15: C

Although the magnitude of acceleration decreases after 5 seconds he is still increasing his velocity. In this case, the velocity is given by the area under the curve. Summing the velocity gained over each second gives the final velocity, with squares here corresponding to 1 m/s² x 1 s = 1 m/s. He only ever loses velocity between 0-0.5 s and 9.5-10 s.

Question 16: C

The radius and tangent to a circle always form a right angle, so using Pythagoras:
3^2 m^2 + 4^2 m^2 = X^2 m^2
X = 5 m

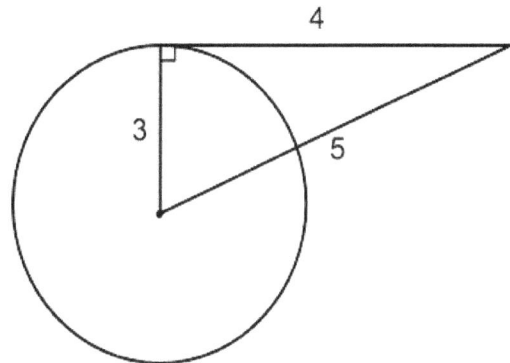

Question 17: A

DNA consists of 4 bases: adenine, guanine, thymine and cysteine. The sugar backbone consists of deoxyribose, hence the name DNA. DNA is found in the cytoplasm of prokaryotes.

Question 18: F

Mitochondria are responsible for energy production by ATP synthesis. Animal cells do not have a cell wall, only a cell membrane. The endoplasmic reticulum is important in protein synthesis, as this is where the proteins are assembled.

Question 19: F

If you aren't studying A-level biology, this question may stretch you. However, it is possible to reach an answer by process of elimination. Mitochondria are the 'powerhouse' of the cell in aerobic respiration, responsible for cell energy production rather than DNA replication or protein synthesis. As energy producers they are required in muscle cells in large numbers, and in sperm cells to drive the tail responsible for movement. They are enveloped by a double membrane, possibly because they started out as independent prokaryotes engulfed by eukaryotic cells.

Question 20: A

The majority of bacteria are commensals and don't lead to disease.

Question 21: C

Bacteria carry genetic information on plasmids and not in nuclei like animal cells. They don't need meiosis for replication, as they do not require gametes. Bacterial genomes consist of DNA, just like animal cells.

Question 22: C

Active transport requires a transport protein and ATP, as work is being done against an electrochemical gradient. Unlike diffusion, the relative concentrations of the materials being transported aren't important.

Question 23: D

Meiosis produces haploid gametes. This allows for fusion of 2 gametes to reach a full diploid set of chromosomes again in the zygote.

Question 24: B

Each three-block combination is mutually exclusive to any other combination, so the probabilities are added. Each block pick is independent of all other picks, so the probabilities can be multiplied. For this scenario there are three possible combinations:

P(2 red blocks and 1 yellow block) = P(red then red then yellow) + P(red then yellow then red) + P(yellow then red then red) =

$(\frac{12}{20} \times \frac{11}{19} \times \frac{8}{18}) + (\frac{12}{20} \times \frac{8}{19} \times \frac{11}{18}) + (\frac{8}{20} \times \frac{12}{19} \times \frac{11}{18}) =$

$2 \quad \frac{\times 12 \times 11 \times 8}{20 \times 19 \times 18} = \frac{44}{95}$

Question 25: C

Multiply through by 15: $3(3x + 5) + 5(2x - 2) = 18 \times 15$

Thus: $9x + 15 + 10x - 10 = 270$

$9x + 10x = 270 - 15 + 10$

$19x = 265$

$x = 13.95$

Question 26: C

This is a rare case where you need to factorise a complex polynomial:

(3x   )(x   ) = 0, possible pairs: 2 x 10, 10 x 2, 4 x 5, 5 x 4

(3x - 4)(x + 5) = 0

3x - 4 = 0, so x = $\frac{4}{3}$

x + 5 = 0, so x = -5

Question 27: C

$\frac{5(x-4)}{(x+2)(x-4)} + \frac{3(x+2)}{(x+2)(x-4)}$

$= \frac{5x-20+3x+6}{(x+2)(x-4)}$

$= \frac{8x-14}{(x+2)(x-4)}$

Question 28: E

p α $\sqrt[3]{q}$, so p = k $\sqrt[3]{q}$

p = 12 when q = 27 gives 12 = k $\sqrt[3]{27}$, so 12 = 3k and k = 4

so, p = 4 $\sqrt[3]{q}$

Now p = 24:

24 = 4$\sqrt[3]{q}$, so 6 = $\sqrt[3]{q}$ and q = 6^3 = 216

Question 29: A

8 x 9 = 72

8 = (4 x 2) = 2 x 2 x 2

9 = 3 x 3

(2   x 2 x 2 x 3 x 3)^2 = 2 x 2 x 2 x 2 x 2 x 2 x 3 x 3 x 3 x 3 = 2^6 x 3^4

Question 30: C

Note that 1.151 x 2 = 2.302.

Thus: $\frac{2 \times 10^5 + 2 \times 10^2}{10^{10}} = 2 \times 10^{-5} + 2 \times 10^{-8}$

$= 0.00002 + 0.00000002 = 0.00002002$

END OF SECTION

Section 1C

Passage 1

Question 1: C
This is the most accurate definition of a fallacy. Indeed, the first paragraph discusses that the gambler's fallacy is indeed a falsehood.

Question 2: C
The Monte Carlo fallacy and the Gambler's fallacy are labels for the same phenomenon. Accordingly, the events in Monte Carlo are not the definition of the fallacy, but merely why it has its name. Answer C fits in best with the explanation provided in the first paragraphs. It is the fact that there have already been lots of (previous) black rolls that people believe that there will be a red due.

Question 3: D
People were betting before the 26th spin so C is incorrect. E is incorrect as the passage does not indicate that they knew of the gambler's fallacy. A is impossible as it's not possible to know with certainty that a future event will happen and there's no evidence to suggest that B was the case. As explained in the passage, the gambler's fallacy operated to engender a false belief that one's odds of getting an *independent* future result (i.e. a red) increased because of the fact that there were lots of blacks beforehand.

Question 4: D
This can be assumed as the gambler's fallacy as it involves at least two independent events – that because one (or more) previous events occurred in a given way means that it's likely that a future event will occur in a particular way.

Question 5: D
Since the experimental group did not differ from the control group, it is clear that teaching probability alone may not change things. It does not prove A as not all possible methods of alleviating the gambler's fallacy were used in the study. B and C don't follow, and E is explicitly mentioned in the passage.

Question 6: D
The gambler's fallacy appears to operate whether or not one has a belief in it (based on the study in the preceding paragraph) therefore, B and C are incorrect. While A is true, it doesn't explain _why_ those in the first group tended to predict tails more. Also, point A was the same for both groups. E is too vague. Thus, D has to be correct. It also follows from the author's explanation of the gambler's fallacy in the beginning of the passage that that is what was being tested here in the final paragraph.

Passage 2

Question 7: C

The essence of the libertarian argument is that people should do what they want so long as they're exercising 'free will' (according to the passage). Therefore, A, B, D, and E are all incorrect. It doesn't matter if there is some risk to the provider or if it is against nature, as long as people consent, it's fine. Accordingly, when one does not give informed consent, arguably one is not exercising free will. Therefore, the libertarian argument would not be relevant in those circumstances.

Question 8: D

It is explicitly stated in the passage that people in Wales are presumed to consent <u>unless</u> they have indicated otherwise, so it is thus clear that people in Wales still have a choice as to whether to donate their organs.

Question 9: A

The author throughout has based the arguments in the passage on the case for allowing the sale of human organs, so point A is correct.

Question 10: A

Whether the poor are exploited through a system for organ sales is open to debate and arguments on either side can be equally valid. This can't be tested as being true or false. Therefore, it is an opinion. All other options are assertions of fact.

Question 11: D

This is never suggested in the passage – while the author believes that there wouldn't be exploitation of the poor, it is never suggested or implied that the poor should donate their organs as opposed to the rich.

Passage 3

Question 12: E

This is the only option that has to be true in order to sustain the author's argument. The author based his point on the fact that the officer didn't want to speak with Tatchell. However, if the student's unions would have been fine with Tatchell speaking, then the author would, in fact, have a platform to speak.

Question 13: D

It is clear that the author's assertions and arguments point towards Tatchell being allowed to speak but the question was asking what the author's views were as to freedom of speech generally. Option D, thus, better describes the author's main position as adopted throughout the passage.

Question 14: E

Whether such censoring is concerning or not is simply an opinion – some may find it wrong, but others may legitimately find it OK because it might offend people. It cannot be tested as it can't be true or false. On the other hand, all other options are assertions of fact. In regard to B, while the question of 'offensiveness' in itself is an opinion, the statement that "Offensiveness has a number of different meanings' is a factual assertion – we can simply have a look at whether there's more than one meaning of offensiveness or not. Therefore, point B taken as a whole is an assertion of fact.

Question 15: E

This is the only point that logically follows from the passage. It is not suggested that any students are extremists (so A is incorrect), it is not suggested or implied that students generally are against free speech (so B is incorrect), it is not clear that many students are against it (so C is incorrect) and finally, the passage does not assume or imply that student unions express the views of the student body.

Passage 4

Question 16: D

While A is true, it is not the main reason for the US bringing the case against Great Britain. B isn't correct as it's not clear that the seals were the US' and C isn't correct as it's neither stated nor implied in the passage that the Canadian ships breached US waters. The passage explicitly states that the depletion of seal stocks led to the conflict.

Question 17: C

The passage explicitly says that Great Britain's counsel persuaded the tribunal. Therefore, it's clear that Great Britain won the case. Canada was not a party to proceedings as Great Britain was representing them (so B is incorrect).

Question 18: B

This explains why the US' argument was convenient because this means that Canada was acting illegally while the US was acting lawfully and would have supported the US' case. A does not provide an explanation. C is not true and D, in fact, makes the US' argument less convincing. The author did not use the word 'convenient' in the context of E.

Question 19: D

This is true as while both sides could seal in the exclusion zone, the tribunal held that they were both restricted in the extent of their sealing.

Question 20: C

The regulation didn't refer at all to the Pribilof Islands. Accordingly, the US would still be free to seal there as they always had done (so long as there were enough seals there).

Passage 5

Question 21: D

A is obviously incorrect as opponents and supporters disagree on the existing wage rates. B is incorrect – we don't know whether opponents and/or supporters believe that fairness has different meanings in different contexts. The author seems to imply this, but the author does not imply that opponents and supporters think that it can have different meanings. However, the author points out that the supporters have one view of fairness (market-based wage rate) and opponents have another view (based on differences in wages). Therefore, both sides consider fairness and so D is correct. E is obviously incorrect because opponents of footballers' wages do not believe that the current outcome is fair.

Question 22: A

This is a tricky question involving the use of two negatives ('not' and 'inconsistent') in the question. Simply put, which of the following responses doesn't contradict both the opponents and supporters arguments? Opponents and supporters all disagree on B, C, D, and E. Therefore, A has to be correct. Indeed, supporters of footballers' high wages never deny that luck played a part.

Question 23: A

This is because the passage never says that hard work is unimportant or not a factor. It's perfectly logical for hard work to be a factor alongside one's birth talents. B is wrong because it's not obvious – the author says that 'careful' political debate is required to determine whether there should even be reform. C is wrong because two different formulations of fairness are highlighted in the passage. D is wrong because the author states that luck has a role to play. E is obviously incorrect.

Question 24: E

It is clear that this sentence is introducing a contrasting point because of the author's use of the word 'however'. Accordingly, having just brought up the possibility of a wage cap, it means that the author is suggesting some kind of limitation to that in this sentence. E is the only choice that fits in with this.

END OF SECTION

Section 2

Question 1: Design an experiment to deduce the sensitivity of a snake's hearing. Explain everything you would do, and your rationale for doing so.

This question is different to most others you will encounter and does not demand the usual structure. Despite this, there should be a logical flow to the answer, explained comprehensibly and clearly, and reasoning should be justified adequately to provide support.

Introduction:
A brief overview of the independent and dependent variables would be appropriate, as well as a short hypothesis about what you may expect to observe.

Points to consider and justify:
- Independent variable = what you will change, probably the frequency or volume of the sound. You will be required to operationalize this variable, considering the equipment used to provide the sound – such as a computer, and the method of delivery to the snake – through speakers, through headphones etc.
- Dependent variable = what you will measure, to do with the snake's response to the sound. This could be operationalized in a number of ways, such as a detectable movement, which has been previously learnt by association, or simply a twitch – but this should be scientific and objective. Any equipment or materials, including an observer, should be identified.
- Sample and repeats = how many different snakes will you test? Will you repeat the experiment on the same snake a number of times?
- Setting = is this a laboratory experiment or will it be taking place in a naturalistic environment?
- Control variables = it must be clear that you have considered potential confounding factors and have taken steps to control them. These may be extraneous noise (control for by taking place in a soundproof room), errors in determining when the snake has heard the noise (repeats are probably the easiest way of eliminating this), sample bias – are you testing just one breed of snake or lots of different ones? Observer bias – if an observer is detecting movement, ensure they are doing this reliably.
- Interpretation of results – include how you will present your data and suggest any data manipulation techniques (graphs, statistical tests) outlining what you hope this will achieve.
- Reporting the experiment – explain how you might wish to write up the experiment and any further research you may want to highlight.

Ideas for justification:
- Convenience
- Time taken vs accuracy
- Validity
- Reliability
- Access/understanding

Conclusion:
Summarise the experiment in a couple of sentences AND your overall justification about what you wish to achieve, identifying both the positives and potential limitations or need for further study.

Question 2. "The eternal mystery of this world is its comprehensibility"
To what extent is the world comprehensible?

Introduction:

- Include a definition of comprehensible – understandable, intelligible
- You may wish to consider the extent to which comprehensibility in itself is subjective, and therefore your essay will be biased towards your own views on comprehensibility.
- Introduce a summary of the ideas you want to explore within the essay on both sides – arguments for the world being comprehensible and arguments against (see ideas below)

Points for comprehensibility:

- We are restrained by the constraints of our own human mind – is the question how much can the human mind comprehend? The world is limitless until we reach the limits of our own minds.
- We are still discovering more and more about the world and haven't reached our limit yet – there is always more to discover and further our knowledge and understanding. Until we can no longer advance in our scientific understanding, the world is comprehensible and in our hands to discover.
- One could argue that we are the most advanced and evolved organisms in the world and therefore everything we need to understand must be less complicated than ourselves.
- With an understanding of history, we can understand why things are the way they are – as long as we learn from the past, surely there is no more in the future we cannot reach?

Points against comprehensibility:

- One could argue that the world is contained within our minds, and since we are within our minds we cannot objectively understand them – do we have to be outside of the world to comprehend it, and therefore not be of this world?
- The human mind may be so complex that it is, paradoxically not complex enough to understand itself.
- There must be a limit to our understanding, since the world has limits and our brain is a limited place for processing.
- We do not know what else there is to comprehend, and therefore we will never know when we have reached the limits of our comprehensibility – therefore never fully reaching complete understanding.
- Would we be God if we could comprehend the world and everything in it?

Conclusion:

- Include a summary of all points made, and give a balanced overview of both sides of the argument.
- Reference back to the original quote – it may be relevant to emphasise the word 'mystery' in the context.
- Draw together your points in some kind of concluding statement – either to say the world is completely comprehensible or to say that it is only comprehensible to some extent would be justified, so long as this is logically and critically summarised and justified.

3. "The greatest obstacle to learning is education"
Argue for or against this statement.

Introduction:

- Consider a definition for learning – the acquisition of knowledge; versus education – a structured approach to transferring information about the world from one person to the other.
- Identify the ways through which knowledge can be acquired – through teaching, experience, study – emphasise that learning doesn't have to be purposeful, whereas education involves the intention of learning (whether or not it is achieved).
- This essay asks for ONE perspective – introduce the argument you wish to pursue.

Possible arguments for:

- The greatest feats of learning could be considered the acquisition of language and/or movement – all of which are done without and before education.
- Education is directive – based on the views of a few people about what they believe is important –and therefore it narrows our minds to only the concepts taught.
- Everyone learns in different ways – creatively, actively, through vision, through hearing and education, as we know it is too structured to suit everyone.
- School is a bubble – set apart from the real world – you could argue that the real learning happens once children leave school and have to fend for themselves.
- Have homeless people or those with no access to education learnt less? They may have learnt different skills, but broader, more practical ones.
- Learning involves actively engaging with the information given, and requires an understanding – you could say that education teaches simply recall of facts, not true knowledge of the world.
- Examples of people who have achieved much without a formal education e.g. Alan Sugar

Possible arguments against:

- You could say that education is feeding the natural curiosity within human beings – we naturally want to learn more about the world and formally educating people of this is an easy and natural way to pass on knowledge.
- Education provides not just academic knowledge, but social knowledge; how to behave, what is right and what is wrong, how to make friends. There isn't such a concentrated opportunity to learn these things in other places.
- Reading and writing are basic skills required for so many careers – education opens the doors to further learning
- How could one say that a person has learnt nothing after graduating from school? Therefore, is cannot be a barrier to learning, because children surely learn something!
- Knowledge needs to be passed on to those capable of furthering it – and therefore education provides a means of passing on knowledge and developing our understanding of concepts – if education didn't exist, development would be slower, so we must be learning something.
- Uneducated people make up the majority of our unemployment figures and are often more likely to turn to drugs, alcohol and develop other social problems.

Conclusion

This is an argumentative essay, so the conclusion MUST reach a decision. Summarise the key ideas of the essay and dismiss any opposing perspectives.

4. Does a vacuum really exist?

Introduction:

- This must include a definition of vacuum – a space entirely devoid of anything.
- You may wish to clearly indicate what you mean by 'anything' – matter, time, etc.
- This could be considered philosophically or physically – make it clear which (or both) perspective you are wishing to take.
- Introduce the key ideas on both sides you wish to pursue on both sides of the argument – reasons for it existing and reasons against (see below)

Arguments for:

- How can we prove that 'nothing' exists – in proving there is nothing, there must be something, in order for there to be something to prove.
- Just because our brains cannot comprehend a place where there is nothing, that does not mean that it doesn't exist – we just may not be adequately equipped to understand it.
- We can never prove that a vacuum doesn't exist, because there is no way of finding it if there is nothing there – so we must assume it exists if we cannot disprove it.
- If there is a place where there is something, there must also be a place where there is nothing, in order for the place where there is something to be valid.

Arguments against:

- Without matter, there is an absence of anything, and therefore without anything, there is nothing – so it cannot exist.
- There are fields and properties and relativity everywhere in the known universe, provided the physics applies equally everywhere, which means there is always something there, so there is nowhere in the universe that there is nothing.
- In order for something to exist, it must have properties which can be defined and demonstrated within the realms of our reasoning – since this is not the case for vacuums, they may not exist.
- Our understanding of physics is by no means complete and therefore our ideas about what matter is and how to define it are likely to change – therefore there is no reason to believe there must be a place where there is nothing – it could just be filled by something we haven't discovered yet.

Conclusion:

- Draw together all the points made on both sides of the argument – summarise the key points and bring them down to earth again.
- The conclusion must be solid and clear – even if not particularly complicated, it must be logical and understandable, in order to bring together this potentially confusing subject.
- You may wish to reach a decision, or decide that it depends on the perspective you take – as long as this is correctly justified.

END OF PAPER

Final Advice

Arrive well rested, well fed and well hydrated

The PBSAA is an intensive test, so make sure you're ready for it. Unlike the UKCAT, you'll have to sit this at a fixed time (normally at 9AM). Thus, ensure you get a good night's sleep before the exam (there is little point cramming) and don't miss breakfast. If you're taking water into the exam, then make sure you've been to the toilet before so you don't have to leave during the exam. Make sure you're well rested and fed in order to be at your best!

Move on

If you're struggling, move on. Every question has equal weighting and there is no negative marking. In the time it takes to answer on hard question, you could gain three times the marks by answering the easier ones. Be smart to score points- especially in section 2 where some questions are far easier than others.

Make Notes on your Essay

You may be asked questions on your PBSAA essay at a University interview. Sometimes you may have the interview as late as March which means that you **MUST** make short notes on the essay title and your main arguments after the essay.

Afterword

Remember that the route to a high score is your approach and practice. Don't fall into the trap that "*you can't prepare for the PBSAA*"– this could not be further from the truth. With knowledge of the test, some useful time-saving techniques and plenty of practice you can dramatically boost your score.

Work hard, never give up and do yourself justice.

Good luck!

Acknowledgements

I would like to express my sincerest thanks to the many people who helped make this book possible, especially the 5 Oxbridge Tutors who shared their expertise in compiling the huge number of questions and answers.

Rohan

About Us

Infinity Books is the publishing division of *Infinity Education Ltd*. We currently publish over 85 titles across a range of subject areas – covering specialised admissions tests, examination techniques, personal statement guides, plus everything else you need to improve your chances of getting on to competitive courses such as medicine and law, as well as into universities such as Oxford and Cambridge.

Outside of publishing we also operate a highly successful tuition division, called UniAdmissions. This company was founded in 2013 by Dr Rohan Agarwal and Dr David Salt, both Cambridge Medical graduates with several years of tutoring experience. Since then, every year, hundreds of applicants and schools work with us on our programmes. Through the programmes we offer, we deliver expert tuition, exclusive course places, online courses, best-selling textbooks and much more.

With a team of over 1,000 Oxbridge tutors and a proven track record, UniAdmissions have quickly become the UK's number one admissions company.

Visit and engage with us at:
Website (Infinity Books): www.infinitybooks.co.uk
Website (UniAdmissions): www.uniadmissions.co.uk
Facebook: www.facebook.com/uniadmissionsuk
Twitter: @infinitybooks7

www.ingramcontent.com/pod-product-compliance
Lightning Source LLC
Chambersburg PA
CBHW082108210326
41599CB00033B/6627